CREATING A NATIONAL FRAMEWORK FOR CYBERSECURITY: AN ANALYSIS OF ISSUES AND OPTIONS

CREATING A NATIONAL FRAMEWORK FOR CYBERSECURITY: AN ANALYSIS OF ISSUES AND OPTIONS

ERIC A. FISCHER

Nova Science Publishers, Inc.
New York

LIBRARY OF CONGRESS CATALOGING-IN-PUBLICATION DATA

Fischer, Eric A.
Creating a national framework for cybersecurity : an analysis of issues and options / Eric A. Fischer.
p. cm.
ISBN 978-1-60456-559-1 (pbk.)
1. Terrorism--United States--Prevention. 2. Cyberterrorism--United States.--Prevention. 3. Computer security--United States. 4. Computer networks--Security measures--United States. 5. National security--United States. I. Title.
HV6432.F574 2008
363.325'9004--dc22

2008011199

Published by Nova Science Publishers, Inc. ✢ New York

CONTENTS

Preface		**vii**
Chapter 1	Summary	**1**
Chapter 2	What Is Cybersecurity?	**5**
Chapter 3	Where are the Major Weaknesses in Cybersecurity?	**9**
Chapter 4	What are the Major Means of Leverage?	**27**
Chapter 5	What Roles should Government and the Private Sector Play?	**49**
References		**61**
Index		**83**

PREFACE

Even before the terrorist attacks of September 2001, concerns had been rising among security experts about the vulnerabilities to attack of computer systems and associated infrastructure. Yet, despite increasing attention from federal and state governments and international organizations, the defense against attacks on these systems has appeared to be generally fragmented and varying widely in effectiveness. Concerns have grown that what is needed is a national cybersecurity framework — a coordinated, coherent set of public- and private-sector efforts required to ensure an acceptable level of cybersecurity for the nation.

As commonly used, *cybersecurity* refers to three things: measures to protect information technology; the information it contains, processes, and transmits, and associated physical and virtual elements (which together comprise *cyberspace*); the degree of protection resulting from application of those measures; and the associated field of professional endeavor. Virtually any element of cyberspace can be at risk, and the degree of interconnection of those elements can make it difficult to determine the extent of the cybersecurity framework that is needed. Identifying the major weaknesses in U.S. cybersecurity is an area of some controversy. However, some components appear to be sources of potentially significant risk because either major vulnerabilities have been identified or substantial impacts could result from a successful attack. — in particular, components that play critical roles in elements of critical infrastructure, widely used commercial software, organizational governance, and the level of public knowledge and perception about cybersecurity.

There are several options for broadly addressing weaknesses in cybersecurity . They include adopting standards and certification, promulgating best practices and guidelines, using benchmarks and checklists, use of auditing, improving training and education, building security into enterprise architecture, using risk management, and using metrics. These different approaches all have different strengths and weaknesses with respect to how they might contribute to the development of a national framework for cybersecurity. None of them are likely to be widely adopted in the absence of sufficient economic incentives for cybersecurity.

Many observers believe that cyberspace has too many of the properties of a commons for market forces alone to provide those incentives. Also, current federal laws, regulations, and public-private partnerships appear to be much narrower in scope than the policies called for in the *National Strategy to Secure Cyberspace* and similar documents. Some recent laws do provide regulatory incentives for corporate management to address cybersecurity issues. Potential models for additional action include the response to the year-2000 computer problem and federal safety and environmental regulations. Congress might consider encouraging the widespread adoption of cybersecurity standards and best practices, procurement leveraging by the federal government, mandatory reporting of incidents, the use of product liability actions, the development of cybersecurity insurance, and strengthened federal cybersecurity programs in the Department of Homeland Security and elsewhere. This report will be updated in response to significant developments in cybersecurity.

Chapter 1

SUMMARY

Even before the terrorist attacks of September 2001, concerns had been rising among security experts about the vulnerabilities to attack of computer systems and associated infrastructure. There were several reasons for those rising concerns. First, computer systems were becoming increasingly powerful and increasingly interconnected, with many enterprises in the public and private sectors coming to rely on them for fundamental business functions. Second, the size and reach of the Internet was growing dramatically. Not only were more and more businesses and households in the United States using the Internet, but the same phenomenon was happening worldwide. Third, the number and sophistication of attacks by criminals and vandals was growing, and many experts thought that terrorists and other adversaries were preparing to launch attacks on computer systems via the Internet or other means. Those trends have generally continued over the last several years.

Yet, despite increasing attention from federal and state governments and international organizations, the defense against attacks on these systems has appeared to be generally fragmented and varying widely in effectiveness. Even with the establishment of the Department of Homeland Security by the Homeland Security Act of 2002 (P.L. 107-296), with its consolidation of several cybersecurity efforts within the Information Assurance and Infrastructure Protection Directorate, and the subsequent publication of the *National Strategy to Secure Cyberspace (NSSC),*[1] concerns grew that a more coordinated, coherent approach — what might be called a national cybersecurity framework[2] — was needed. What such a framework should consist of, whom it should apply to, and how it should be developed and implemented have remained uncertain. Several processes are underway that may contribute to the development of such a framework, ranging from some sector specific activities to proposals for federal legislation. The issues associated with that development can be difficult to understand and address for several reasons, perhaps most notably because of the sheer size, complexity, and interconnectedness of the information infrastructure and associated technology and applications. The purpose of this report is to lend structure to the debate about those issues by examining some fundamental concepts and questions relating to a framework.

A national cybersecurity framework can be thought of as the essential set of public- and private-sector efforts required to ensure an acceptable level of cybersecurity for the nation. To be effective, such a framework would need to operate in at least four dimensions. One, perhaps the most obvious, consists of the elements of cybersecurity. It includes both the general approach — e.g., goals, best practices, benchmarks, standards — and specific areas of focus, such as technology, process, and people. A second dimension is the components of cyberspace — what would be covered by the framework. That includes both specific elements, such as computer operating systems and Internet servers, and the sectors which would be involved. A third dimension is the method of application. For example, should the framework be required, voluntary, or ad hoc? The fourth dimension is the functions and goals of the framework. Is its purpose to defend against crime, to improve the environment for electronic commerce, to protect critical infrastructure, or some combination of those?

No consensus proposal for a cybersecurity framework has yet emerged, and suggestions tend to focus on different approaches and components. Some of those emphasize cybersecurity policies and goals, others procedures, still others technology. Some stress standards, others best practices or benchmarks, and still others focus on guidelines. This diversity of possible approaches can complicate examination of the issues. A further complication may arise from the lack of consensus meanings for terms used to denote different approaches.

To examine what kind of framework may be needed and how it might be implemented, it may be helpful to address three questions:

1. *Where are the major cybersecurity weaknesses currently, and where might weaknesses be anticipated in the future?* The term *weaknesses* as used here includes vulnerabilities and associated risks as those terms are usually understood, but also other factors that might negatively impact cybersecurity but might not usually be considered vulnerabilities or risks. For example, misperceptions about risks might be a weakness. A weakness is *major* if failure to address it could realistically have a significant national impact on the economy, public safety, or other critical

services. The assessment of weaknesses will also determine the goals of a framework to a significant extent.

2. *What are the major means of leverage for addressing those weaknesses?* These could include such approaches as the adoption of standards or best practices, improvements in software engineering, investment in training and education, or correction of market failures.
3. What roles should government and the private sector play in the use of those means of leverage to address current and potential future weaknesses? It might be, for example, that market forces are sufficient to address the concerns. Alternatively, incentives might be needed to promote voluntary measures, or regulation might be required. Among the policy options that Congress could consider are encouraging broader use of cybersecurity standards and best practices in the private sector, using federal procurement practices to leverage general improvements in products and services, encouraging mandatory reporting of security incidents, facilitating product-liability actions in response to inadequate cybersecurity practices, encouraging the development of cybersecurity insurance, and strengthening federal cybersecurity programs.

This report addresses each of those questions in turn. However, before doing so, it may be useful to discuss exactly what the term cybersecurity refers to.

Chapter 2

WHAT IS CYBERSECURITY?

One of the prerequisites for developing a common national framework for cybersecurity is a common understanding of what this and related terms mean. Achieving that can be difficult, for several reasons. Perhaps the major one is complexity. There are many components of cyberspace and many potential components of a framework. A variety of stakeholders will be involved with, exposed to, and in some cases predisposed to focus on different parts of cyberspace, different elements of a framework, and different approaches to security. Consequently, attempts to create a coordinated national framework could be challenging.

Another problem is that there appears to be no generally accepted definition of cybersecurity, and several different terms are in use that have related meanings. For example, *information security* is defined in some subsections of federal copyright law to mean "activities carried out in order to identify and address the vulnerabilities of a government computer, computer system, or computer network" (17 U.S.C. 1201(e), 1202(d)), and, in the Federal Information Security Management Act (FISMA, P.L. 107-296, Title X, 44 U.S.C. 3532) as "protecting information and information systems from unauthorized access, use, disclosure, disruption, modification, or destruction."

The term *information assurance* (IA) is also used. One section of federal military law defines it to include computer and network security as well as any other information technology so designated by the Secretary of Defense (10 U.S.C. 2200(e)). The National Security Agency (NSA) defines information assurance as

> Measures that protect and defend information and information systems by ensuring their availability, integrity, authentication, confidentiality, and no repudiation. These measures include providing for restoration of information systems by incorporating protection, detection, and reaction capabilities.[3]

Assurance more generally refers to the level of confidence in the effectiveness of security.[4] Also, both information security and information assurance may not be limited to electronic systems but may refer more broadly to the protection of information or data in whatever format or medium it exists.

In the context of financial services, *electronic security* or *e-security* has been defined as follows:

> E-security can be described on the one hand as those policies, guidelines, processes, and actions needed to enable electronic transactions to be carried out with a minimum risk of breach, intrusion, or theft. On the other hand, e-security is any tool, technique, or process used to protect a system's information assets....E-security enhances or adds value to an unprotected network, and is composed of both a "soft" and a "hard" infrastructure. Soft infrastructure components are those policies, processes, protocols, and guidelines that create the protective environment to keep the system and the data from compromise. The hard infrastructure consists of the actual hardware and software needed to protect the system and its data from external and internal threats to security.[5]

Neither federal law nor the *NSSC* define *cybersecurity*, and the latter uses the term interchangeably with "cyberspace security." The implication, presumably, is that the former term is shorthand for the latter, which is also not defined. However, in general usage, *cyberspace* is more of a metaphor than a precise concept, and it has different meanings in different contexts. The *NSSC* uses it to refer to computers and the hardware connecting them.[6] In common parlance, it is often used somewhat differently, referring to a kind of virtual space, created by computer networks, within which people and computers perform various activities such as email, financial transactions, data processing, and system control.[7] In this report, *cyberspace* means the combination of the virtual structure,[8] the physical components that support it, the information it contains, and the flow of that information within it.

A cybersecurity bill introduced in the 108th Congress, the Department of Homeland Security Cybersecurity Enhancement Act — H.R. 5068/Thornberry; reintroduced in the 109th Congress as H.R. 285 — defines cybersecurity as

> ...the prevention of damage to, the protection of, and the restoration of computers, electronic communications systems, electronic communication services, wire communication, and electronic communication, including information contained therein, to ensure its availability, integrity, authentication, confidentiality, and nonrepudiation.[9]

This proposed legislative definition is similar to the NSA definition of information assurance described above, but with a greater stress on communications systems.[10] Also, this definition emphasizes outcomes — prevention, protection, and restoration — rather than processes to achieve those outcomes.

The potential fuzziness of the term *cybersecurity* could be a problem in the context of developing a national framework to the extent that it impacts the ability of different stakeholders to reach agreement on elements of the framework. However, because information technology and cyberspace itself continue to evolve rapidly, a rigid definition would likely lose its utility quickly. Keeping the concept as flexible as possible may be beneficial.

As it is commonly used, cybersecurity appears to refer to three things:

1. A set of activities and other measures intended to protect — from attack, disruption, or other threats — computers, computer networks, related hardware and devices software, and the information they contain and communicate, including software and data, as well as other elements of cyberspace.[11] The activities can include security audits, patch management,[12] authentication procedures, access management, and so forth. They can involve, for example, examining and evaluating the strengths and vulnerabilities of the hardware and software used in the country's political and economic electronic infrastructure. They also involve detection and reaction to security events, mitigation of impacts, and recovery of affected components. Other measures can include such things as hardware and software firewalls, physical security such as hardened facilities, and personnel training and responsibilities.
2. The state or quality of being protected from such threats.[13]
3. The broad field of endeavor, including research and analysis, aimed at implementing and improving those activities and quality.[14]

The kinds of attack or disruption contemplated are generally those originating with humans — vandals, criminals, terrorists, nation-states — but other sources are possible, such as accidents, major weather events, or earthquakes.

This three-part sense is how cybersecurity is used in this report. In this usage, information security, information assurance, e-security, and even cyberspace security (as used in the *NSSC*) are aspects of, but not synonymous with, cybersecurity.

Chapter 3

WHERE ARE THE MAJOR WEAKNESSES IN CYBERSECURITY?

Cyberspace is large, somewhat amorphous, and growing. It is interconnected in ways that can be difficult to characterize or even identify. It is also global, and most of it is in the private sector. Therefore, a thorough determination of what parts of and activities in cyberspace should, and even can, be involved in a national framework for cybersecurity is difficult to do. The complexity of cyberspace and its components, even within organizations, makes it both difficult to test and to predict how systems will behave under unusual circumstances, such as might arise from an unanticipated cyberattack.[15] However, the issue can be made more tractable by first identifying those components which, if successfully attacked, would yield damage that would be considered unacceptable to government and/or the public. From this perspective, there appear to be three classes of attack to consider:

- *Service Disruption.* Those which cause a loss of service, such as by making unavailable part of cyberspace or activities that depend on it. This could include, for example, a loss of emergency or transportation communication systems, or an extended unavailability of electronic financial transactions or utilities such as electricity. This unavailability could result from disabling of networks through a variety of attacks such as denial of service (DoS), corruption or destruction of information such as financial records, or destruction of physical infrastructure such as components of the Internet backbone. Disruptions may be limited to a particular organization, region, or sector, or they could be broader and even global in scope.

- *Theft of Assets.* Those which involve theft or other appropriation and subsequent misuse of critical information on a large enough scale to have major impact, such as on financial markets.
- *Capture and Control.* Those which involve taking control of components of cyberspace and using them as weapons against other critical activities or elements of infrastructure, such as using compromised home computers to launch DoS attacks against one or more targets.[16]

Three main channels of attack also exist — *through cyberspace*, such as via worms or other malware,[17] by direct *destruction or alteration of physical structure*, such as buildings or telecommunications lines, or through intentional or inadvertent *actions by a trusted insider*. These channels are not mutually exclusive, and combination attacks are also possible. Because cyberspace is constantly under attack, albeit most often at a low level and largely through the first channel, and because most attacks produce damage which is either minimal or considered acceptable (e.g., as a cost of doing business) by those attacked, a higher threshold of impact might need to be reached before significant efforts at developing a national cybersecurity framework would be considered worthwhile by many. Determining that threshold, in particular the threshold for government action, is not straightforward, especially given that most publicly known cyberattacks to date fall into either a nuisance or vandal category, such as defacing websites, or financial crime, such as stolen credit card or personal identity information. Additionally, the costs of cyberattacks may be difficult to quantify, further complicating the determination of that threshold.[18] In the absence of a significant terrorist cyberattack, the role that government should play in ensuring a specified national level of cybersecurity has been somewhat controversial. Yet, policymakers are also keenly aware that before September 2001 most observers regarded a major attack by foreign terrorists inside the United States an unlikely event. In that context, to identify areas worthy of particular attention, it could be useful to discuss what parts of cyberspace might be in particular need of protection.

WHAT COMPONENTS OF CYBERSPACE ARE AT RISK?

Cyberspace, as used in this report, comprises a huge range of elements arrayed worldwide. Given its size and complexity, a reasonable question is whether a cybersecurity framework is needed for all of cyberspace — potentially a daunting task — or only for certain critical components that are especially important or especially at risk. Cyberspace consists not only of the Internet and computers connected to it, but also any electronic system or device that is or can be connected either directly to the Internet or indirectly through some other device or system, as well as the mechanisms that connect them. These may include such things as automatic teller machines, industrial control systems known as SCADAs,[19] and even telephone and other telecommunications systems. These connections may be obvious, or they may not. Thus, not only is a Web-enabled cellular telephone part of cyberspace, but so is a desktop phone, not only because it is part of the same worldwide telephone system as the cellular phone, but also because that telephone system increasingly relies on computers and the Internet to help manage traffic and for other purposes. Even a computer with no connection to the Internet is part of cyberspace if it has a way of communicating with other computers — such as through floppy disks or other removable media.

Cyberspace also includes the software that runs computers and their connections. It includes the data stored on or generated by those computers and other devices and the transmission of those data to other computers and devices. It includes cables, routers, servers, networks, the Internet backbone, and even satellites used in Internet transmissions. It even has its own atlases[20] and sophisticated electronic mapping techniques to help manage networks and Internet communications. These show that even the virtual dimension of cyberspace is highly structured, and often in ways that may not be obvious.[21]

In addition to the components of cyberspace per se, there are supplementary elements that can be critical with respect to cybersecurity. Perhaps most notable are buildings and other structures within which the physical components of cyberspace are contained, and people with access to cyberspace. An effective cybersecurity framework needs to take such elements into account.

It might be useful to consider the various physical components associated with cyberspace as *cyberspace infrastructure*. This can be conveniently categorized into four segments: Internet hardware, telecommunications infrastructure, embedded computing devices such as control systems, and dedicated computing devices such as desktop computers.[22] Virtual cyberspace — the electronic information that is stored in and flows through the physical components — could be called *cyberspace superstructure.*

Virtually any element of cyberspace can, at least in theory, pose some level of cybersecurity risk, which is generally thought of as a combined assessment of threat, vulnerability, and impact[23] that gives a measure of the overall potential for harm from a vulnerability if no corrective action is taken. *Threat* can have several different meanings, but in this report it refers to a possible attack — for example, the threat of a denial-of-service attack. Descriptions of threats often include both the nature of the possible attack and those who might perpetrate it, as well as the capabilities of potential attackers, and may include some description of the possible consequences if the attack is successful. *Vulnerability* usually refers to a weakness that an attack might exploit — how an attack could be accomplished. Analysis of threats and vulnerabilities, when combined, can lead to an assessment of *risk*. Statements of risk often combine both the probability of a successful attack and some measure of its likely *impact* — the nature and magnitude of economic and other outcomes of a successful attack.

Managing risks may involve several kinds of activities. *Defense* refers to how a system is protected from attack and is often discussed in terms of *countermeasures* or *controls*. Elements may include *prevention* 2 which involves reducing vulnerabilities and implementing other measures to deter attacks; *detection,* which involves identifying and characterizing an attempted attack, either as it occurs or afterward; and *countering* (sometimes also called response), which involves taking corrective measures in response to an attack to stop it or reduce its impact. *Response and recovery* refer to how, and how well, damage is mitigated and repaired and information and functionality are recovered in the event of a successful attack.

Risks are often characterized qualitatively as high, medium, or low.[24] The level of risk varies among different components of cyberspace, and some may therefore deserve more attention than others in the development of an effective framework. Some components are considered to be particularly vulnerable, some are viewed by different groups of attackers as particularly tempting targets, and some would, if compromised, have particularly large impacts. These may not, however, all translate into high risk. For example, a target could be highly vulnerable but under little threat and with a very limited impact resulting from any successful attack. In contrast, a moderately vulnerable target under moderate threat with moderate impact from a successful attack could easily be assessed as being at much higher risk.

Identifying what are the major weaknesses in U.S. cybersecurity is an area of some controversy. While there seems to be general agreement on some problems — such as software vulnerability and the increasing levels of cybercrime — others have in fact remained controversial. Even the question of how much of a concern cyberterrorism (as opposed to cybercrime[25]) should be has been a matter of some dispute. However, terrorists may also engage in cybercrime such as theft, fraud, extortion, or money laundering to finance their efforts. There also appear to be increasing concerns among some observers about the possibility of a growing nexus among hackers, organized crime, and terrorists.[26] Therefore, separating consideration of those activities in discussions of cybersecurity might not be appropriate, at least in some cases.

There appear to be certain candidate components of cyberspace and associated activities that are sources of potentially significant risk because either major vulnerabilities have been identified or substantial impacts could result from a successful attack. They are

- *Components that play critical roles in elements of critical infrastructure.* This could include, for example, computer control systems such as SCADAs used in the chemical and energy industries, and the Internet infrastructure. Another example is information held by financial services industries that could be stolen electronically or otherwise compromised.
- *Software.* In particular, widely used computer programs such as operating systems can be vulnerable to various forms of compromise resulting, for example, in information theft or use of the compromised system as a weapon of attack. This kind of vulnerability has perhaps received more public attention than any other, given that it affects virtually all owners and users of desktop systems.
- *Cybersecurity governance.* Many observers have expressed concerns that corporations and other organizations, including some involved in critical infrastructure sectors (see below), have not developed governance mechanisms sufficiently responsive to cybersecurity needs. Weaknesses have been cited with respect to several aspects of cybersecurity governance, including policies, procedures, and personnel management.
- *Public knowledge and perception.* Observers who have expressed concern about the risk of major cyberattacks from terrorists or other criminals have in many cases pointed to a lack of public awareness about the risk as a weakness, both with respect to lack of knowledge about the steps individuals need to take to defend against attacks and the need for national public- and private-sector effort.

While other potential weaknesses could be identified — for example, security of current Internet protocols, emergency communications, or buildings housing key Internet servers or central exchanges known as peering points — discussion here will be limited to the four cases above, because of their relevance to the issue of a national framework.

CYBERSPACE AND CRITICAL INFRASTRUCTURE

It is obvious from even a cursory examination of cyberspace that it is probably impossible and certainly impractical to secure all of it — if for no other reason than its global nature. But even within the United States, the complexities are daunting. One of the fundamental tenets of cybersecurity is that a simple system is much easier to secure than a complex one — and cyberspace is extraordinarily complex. Determining the elements of cyberspace that should be the focus of cybersecurity is therefore of fundamental importance. One set of components for which there already appears to be general agreement to include consists of those associated with the nation's critical infrastructure.

Some components of cyberspace are also components of the U.S. critical infrastructure (CI),[27] defined as "systems and assets, whether physical or virtual, so vital to the United States that the incapacity or destruction of such systems and assets would have a debilitating impact on security, national economic security, national public health or safety, or any combination of those matters."[28] The *National Strategy for Homeland Security* identified thirteen CI sectors,[29] which DHS has categorized as follows[30]:

- *production industries:* energy, chemical, defense industrial base;
- *service industries:* banking and finance, transportation, postal and shipping;
- *sustenance and health:* agriculture, food, water, public health;
- *federal and state:* government, emergency services;
- *IT and cyber:* information and telecommunications.

Disruption of CI components by natural or anthropogenic events can have significant economic and social impacts. Those impacts can reverberate well beyond the affected industry, as the August 2003 electricity blackout in the northeastern United States demonstrated.

Many CI industries are increasingly dependent on cyberspace, and adequate cybersecurity for those industries is important not only to them, but to other industries, government, and the public. Some examples of components of CI cyberspace that have received particular attention with respect to risk are described below.

Control Systems. Computer systems are often used to control various industrial processes. In many instances, those systems are connected directly or indirectly to the Internet. Their potential vulnerabilities are particularly a concern in industries that are considered part of the nation's critical infrastructure —

especially, energy generation and transportation industry such as electric utilities, oil refineries, and pipeline companies; water utilities; telephone companies; and the chemical industry.[31] The August 2003 electrical blackout in the Northeast has been attributed in part to failure of a computer-controlled alarm system, although it appears to have been caused by malfunctions, not a cyberattack.[32] However, in January of the same year, infection by a computer worm caused a monitoring system to become disabled in an off-line nuclear power plant.[33] In perhaps the best-known example of a cyberattack on control systems, in 2000 a hacker in Australia caused a computerized waste-management system to dump millions of gallons of raw sewage into rivers and parks.[34]

Databases Containing Sensitive Information. Many databases on government and private-sector computer systems contain sensitive information. That can include personal data such as medical records, financial information such as credit card numbers, proprietary business information such as business plans or customer data, security-related data such as risk assessments, and a wide range of other information that might be of interest to competitors, criminals, and terrorists.

Losses from electronic theft and other forms of cybercrime are thought to be in the tens to hundreds of millions of dollars annually in the United States and much larger worldwide. According to some reports, more than half of electronic attacks are directed at financial institutions.[35] However, estimates of losses vary because, among other reasons, institutions are reluctant to share such information because of potential additional losses that could result from damage to the institution's reputation should the information become publicly known.[36]

Competitive pressures often motivate organizations to adopt new information technologies. However, these technologies may also create vulnerabilities by facilitating "more efficient and quicker ways to commit old crimes such as fraud and theft....Disturbingly, as the technology becomes more complex, a perpetrator needs fewer skills to commit these crimes."[37] For example, many companies increasingly use wireless communications ("WiFi" — for wireless fidelity) for networking and other communications. Many of those systems are notoriously vulnerable to compromise by hackers, who can steal passwords and other information or even take over control of the network.[38] The Australian hacker discussed above used a radio to access the sewage control system. While WiFi vulnerabilities can be greatly reduced through application of appropriate security measures, the process can be complex and difficult for many users to implement.

Voting Systems. State and local government are categorized as a CI sector, and like other sectors, they rely increasingly on information technology to provide crucial services. One example is voting systems. Four out of five American voters now cast ballots using systems that rely on computers for casting, counting, or both.[39] While not generally considered part of critical infrastructure, voting systems are central to the functioning of government. Concerns have been raised by many computer security experts about the vulnerabilities of current computer-assisted voting systems to compromise that could change the outcome of an election.[40]

While a focus on cybersecurity for critical infrastructure per se is clearly important, other cyberspace components are also relevant. For example, the education sector, is not generally considered a CI sector, but attacks on components such as institutions of higher learning with significant research programs could have significant impact. Also, CI sectors are largely thought of as being geographically limited to the United States, but cyberspace is global. That means both that attacks outside the United States could have significant impacts within the country, and that the generation of attacks from outside the United States can be of significant concern. Third, cyberspace components that are clearly not part of CI, such as home computers, may be used in attacks.

Software Design Weaknesses

The security problems of much widely used computer software are among the best known cybersecurity weaknesses, because they affect so many computers in homes and businesses. Among these weaknesses, the vulnerabilities of computer operating systems and email programs are among the most widely reported and exploited. They can permit individual computer systems to be probed or even taken over by attackers, with impacts ranging from vandalism to theft to loss of service for a company or a larger segment of cyberspace users. For example, in 2003, computer worms[41] that exploited vulnerabilities in Microsoft Windows operating systems led to disruptions in automatic teller machines and even, in one instance, emergency 911 service, simply by rapidly replicating and propagating themselves, thereby overwhelming computer networks worldwide. The disruptions to financial and emergency services were apparently not anticipated by security experts, nor, probably, by the authors of the worms.[42]

The design of software can have a significant effect on its vulnerability to malware.[43] Both the complexity of the code and the way it is designed can have an impact. It is a general principle of computer security that the more complex a piece of software is, the more vulnerable it is to attack. That is because more complex code will have more places that malware can be hidden and more potential vulnerabilities that could be exploited, and is more difficult to analyze for security problems. In fact, attackers often discover and exploit vulnerabilities that were unknown to the developer.

Software code that is not well-designed from a security perspective is more likely than well-designed code to have weaknesses that could be exploited, as well as places for malware to be hidden. Furthermore, many experts argue that it is impossible with current engineering methods to anticipate all possible weaknesses and points of attack for complex software. However, code can be designed so as to minimize such vulnerabilities, and well-developed procedures have been established to accomplish this goal.[44] Some of those procedures can even be applied to older, legacy systems. Good security design involves not only the code itself, but also the process by which it is developed and evaluated.

Until recently, widely used software was not, for the most part, developed with security as a major goal. That was at least in part because it was not clear that, in the absence of significant breaches, consumers would pay for the extra cost that can be involved in developing more secure software.

Some experts believe that publicly disclosed or *open-source* software provides superior security to proprietary or *closed-source* code.[45] Such experts argue that open-source software is more secure because the open review process is more thorough and can identify more potential security flaws than is possible with proprietary code. Advocates of closed-source code argue, in contrast, that proprietary code makes potential flaws more difficult to discover and therefore to exploit, and that it improves security by providing more control over the personnel, technology, and processes involved in development and maintenance of the code. Since malware has been created for open-source as well as closed-source systems, and since hackers are generally expected to focus on more popular systems, which are currently closed source, the relative security strengths and weaknesses of the two approaches have not been firmly established. However, approaches to improving security that could be applied broadly to different kinds of software would likely be beneficial.

Problems with Organizational Governance

Many observers have expressed concerns that corporations and other organizations have not developed sufficiently responsive governance mechanisms to address cybersecurity needs. It is generally accepted that sound cybersecurity involves a focus on three elements: technology, operations, and personnel.[46] Successful implementation of all three elements requires active involvement by those involved in the governance of an organization.

- The *technology* component focuses on the development, acquisition, and implementation of hardware and software. Organizations have often been criticized for focusing too heavily on this component — the perfect technology, like the perfect lock, is an attractive but elusive security goal.
- The *operations* component focuses on policies and procedures, including such processes as certification, access controls, management, and assessments.
- The *personnel* component focuses on a clear commitment to security by an organization's leadership, assignment of appropriate roles and responsibilities, implementation of physical and personnel security measures to control and monitor access, training that is appropriate for the level of access and responsibility, and accountability.

A focus that is not properly balanced among the three elements may create vulnerabilities. Thus, even an excellent security technology will be minimally effective if it is not properly implemented and used, which requires appropriate governance mechanisms throughout the organization.

Key Aspects of Governance

Weaknesses have been cited with respect to several aspects of cybersecurity governance. Generally, effective governance for cybersecurity would be expected to involve establishing clear and measurable goals, strategies for achieving those goals, and policies and procedures to implement those strategies. These would involve not only operations but personnel management, including the establishment of appropriate roles and responsibilities and accountability for them throughout the organization, as well as recruitment and training. These aspects of governance are discussed in some detail below, because governance is considered by many to be among the most important and complex weaknesses to address.

Goals

Any meaningful framework for cybersecurity should arguably include a clear description of its goals — the desired results or state. Different kinds of goals might be set, and some would likely be more useful than others. For example, a goal might focus only on limiting the number and kinds of attacks that occur. This would have the benefit of being highly tangible and is clearly a desired state, but the relationship between this goal and the other components of the framework may be difficult to determine. Since the number of attempted attacks is determined to a significant extent by the attackers, a low rate of attack does not necessarily reflect effective security. In addition, this kind of goal could create perverse incentives, since, for example, attacks might be reduced by limiting connectivity or computing power, which would often be counterproductive. That is not to say that such a goal is not valuable, but rather that it must be properly developed and characterized.

One well-established approach is to identify *functional goals*, such as those that relate to maintenance of a particular level of operation or performance as opposed to those that focus on prevention of attacks or protection of systems. Three functional goals are commonly described for information security — integrity, availability, and confidentiality.[47] Additional goals — such as accountability, authenticity, reliability, and nonrepudiation — may be added to these basic ones for some applications.[48] A broader functional goal commonly discussed is trustworthiness, which has been defined as

> assurance that a system deserves to be trusted — that it will perform as expected despite environmental disruptions, human and operator error, hostile attacks, and design and implementation errors. Trustworthy systems reinforce the belief that they will continue to produce expected behavior and will not be susceptible to subversion.[49]

Such functional goals have the benefit of being applicable not only with respect to cyberattacks, but also to other sources of potential disruption such as weather events. However, the goals are complex, and it may be difficult to find ways to map them on to other elements of a framework.[50]

It is useful to distinguish goals at different levels — national, sectoral, and organization-specific. The *NSSC* outlines three strategic objectives for the nation:

> Prevent cyber attacks against America's critical infrastructures; Reduce national vulnerability to cyber attacks; and Minimize damage and recovery time from cyber attacks that do occur.[51]

Those objectives, while clearly desirable in concept, may be criticized as somewhat vague, and it is not clear how to determine whether they have been successfully met. Measuring prevention of cyberattacks suffers from the difficulties mentioned in the previous paragraph. *It is not clear how much of a reduction in vulnerability would indicate success, or what it means to "minimize damage and recovery time."* The *NSSC* recommends a set of actions but does not provide a roadmap or other mechanism for assessing those actions against the goals. Presumably, such linkages would be made through subsequent work by federal agencies and the private sector. In general, however, national goals and objectives will of necessity be broad in nature and might best be considered concepts that individual sectors and organizations can use to help them develop more specific goals and objectives.

Sector-specific goals will necessarily vary depending on the mission and focus of the sector. For example, federal goals must be responsive to unique governmental security needs and requirements. Chemical-sector goals would address security of process controls and physical plants, among other things.[52] Organization-specific goals would be expected to be more closely tailored to the individual requirements of each organization.

Whatever focus they take, a set of effective goals should arguably have the following characteristics, among others:

- *Progress toward the goals should be measurable in a meaningful way.* It should be possible to determine to what extent the goals have been met.
- *The goals should provide a basis for appropriate incentives.* They should stimulate improvements in cybersecurity but avoid providing perverse incentives such as inhibiting replacement of obsolete technology.
- *The goals should provide a clear basis for other elements of the framework.* It should be possible to determine how a given element relates to one or more goals.

Strategies

Generally speaking, *strategies* is used as a relatively high-level term referring to a broad set of plans or approaches for meeting cybersecurity goals. For example, the *NSSC* lays out a set of six principles and five largely programmatic priorities for achieving the objectives presented in the document. In this case, the document itself is called a strategy, and is part of a set relating to homeland security and combating terrorism developed by the Bush administration.[53]

The National Security Agency has developed an information assurance strategy called *defense-in-depth* (DID),[54] which focuses on the three elements mentioned above — personnel, technology, and operations — and lays out a set of principles and practices for them. This strategy emphasizes the concept of layered defense. Nonfederal entities have also developed cybersecurity strategy documents — for example, industry groups, corporations, and international organizations.

Because of the various meanings given to *strategy* in different contexts, it is difficult to identify any one set of desired characteristics for the strategic components of a cybersecurity framework. However, among those characteristics are likely to be the following:[55]

- provide an overall methodology for meeting all cybersecurity goals;
- identify resources needs and sources;
- identify organizational responsibilities and roles;
- include ways of measuring progress toward the goals and responding to results of those measurements;
- provide for flexibility and adaptation to changing conditions.

Principles

Many discussions of cybersecurity include statements or lists of principles, which can be thought of as generally accepted characteristics or expectations. While some common themes appear in different descriptions of principles, they generally seem to be developed for specific applications. Among the more widely-cited is published a set of "generally accepted system security principles" (GASSP) that NIST published in 1996. They stress

- the role of computer security in the mission and management of an organization,
- the importance of cost-effectiveness,
- the responsibilities of system owners beyond their own organizations,
- the importance of explicit responsibilities and accountability and of a comprehensive and integrated approach,
- the need for periodic reassessment, and
- the limitations imposed by societal factors such as the need for privacy.[56]

The principles contained in the *NSSC* are somewhat different, in keeping with its broader scope. They contain some similarities to the GASSP but also stress the need for a national effort, a preference for the reliance on market forces rather than government regulation, and the importance of flexibility and multiyear planning.[57] The Information Systems Security Association (ISSA) has been developing a set of Generally Accepted Information Security Principles (GAISP), but these are in effect guidelines (see below).

Among the principles developed for specific application are those laid out in the cybersecurity strategy of the U.S. chemicals sector. They include the importance of involvement of top management, the need for customized solutions, the importance of national and international harmonization, the need for an evolving strategy, and the importance of inclusive participation.[58] The *DID* principles focus on the technology leg of the strategy, stressing layered defense in multiple, customization of protection based on the asset being protected and the threat, robust encryption key management and infrastructure, and intrusion detection infrastructure.

A delineation of principles can be an important component of any approach to cybersecurity governance, providing context for the other components (such as the chemical sector's "integration…with…the global economy") as well as cross-cutting themes (such as NIST's "need for periodic reassessment") and limitations (such as the *NSSC*'s "importance of protecting privacy and civil liberties"). Among the more common themes in different sets are the key role of organizational leaders, the need for considering the environment beyond the organization itself, the importance of context-specificity and adaptability in response to changing circumstances, and the need to take into account other factors such as cost and privacy.

Policies

A *policy* is essentially a set of rules governing how cybersecurity strategies will be applied. Policies can usefully be thought of in terms of levels. A *mission-level* policy lays out broad direction and guidance for an enterprise. For example, the *NSSC* states,

> It is the policy of the United States to prevent or minimize disruptions to critical information infrastructures and thereby protect the people, the economy, the essential human and government services, and the national security of the United States. Disruptions that do occur should be infrequent, of minimal duration and manageable and cause the least damage possible. The policy requires a continuous effort to secure information systems for critical infrastructure and includes voluntary public-private partnerships involving corporate and nongovernmental organizations.[59]

A *program-level* policy provides rules for a specific program or set of activities.[60] Such policies often include the assets to be protected, goals, organizational responsibilities, and compliance parameters, including penalties. A *system-level* policy provides rules for securing a particular system or subsystem. These policies often are based on technical and risk analyses and vary from system to system, depending on requirements. An *issue-level* policy provides rules for a particular issue or area of concern, such as how to handle email attachments. Such policies often cover objectives, responsibilities, and compliance. This kind of policy is likely to require frequent updating in response to changes in technology and other factors.

The cybersecurity policies of an organization serve to provide guidance in meeting stated goals and can also provide incentives — or remove disincentives — for certain behavior. For example, whether or not employees report suspected security breaches may depend in part on the kind of policy the organization has with respect to them. If the policy does not encourage reporting, then employees may be reluctant to do so because of concerns about potential repercussions. Also, policies often set expectations with respect to resource allocation. If cybersecurity is a high policy priority, then it would ordinarily be expected to be a high budget priority as well. A mismatch between an organization's policy and its behavior may have legal ramifications.[61]

Procedures

Procedures can be thought of as specifications of how to perform specific actions, methodologies, or processes. Ideally, cybersecurity procedures would be designed to implement cybersecurity policies and strategies. They may include, for example, steps to take in configuring networks to minimize the risk of successful intrusions, actions to take when an intrusion occurs (including how to report it), and methods for evaluating potential security risks of prospective employees. Although many procedures may be common across different organizations or even sectors, they will likely in general be the most customized and organization-specific of governance components.

Personnel

The components of cybersecurity governance discussed above apply to personnel as well as operations. According to some observers, people are the most important of the three fundamental elements of cybersecurity. It is they who must implement security policies and procedures and defend against any attacks. If they are not adequately skilled and trained, they may be unable to prevent, detect, and react to security breaches, and they may themselves be more vulnerable to a "social engineering" attack, which involves finding and exploiting weaknesses in how people interact with computer systems.[62] In addition, it can be particularly difficult to defend against attack by an insider, so background checks and other controls to minimize that risk are especially important. These considerations may be even more critical for services that are outsourced, in which case direct control over personnel is substantially reduced. According to some observers, such "trusted insiders" pose the most significant threat to an organization's cybersecurity.[63]

It is generally held that effective governance for cybersecurity requires a strong commitment from an organization's leadership — at the level of the chief executive officer, the board of directors, or the equivalent. This may be especially important because returns on investment in security may be difficult to measure.[64] The lack of a clear return on investment may create pressures to under invest. Some evidence suggests that such underinvestment is an issue for many organizations.[65] This may be especially true for small to mid-sized private-sector entities and for state and local government agencies.

If roles, responsibilities, and accountability are not clear and appropriate — which might be the case, if, for example, an organization has no overall cybersecurity policy structure — that can create significant vulnerabilities. A classic case is where responsibilities are too widely distributed; workers may assume that an issue is being addressed by others who share the responsibilities, with the result being that the issue is not properly addressed by anyone.

Another area of concern with respect to personnel is training. Development of a proficient cybersecurity workforce in the United States is listed as a priority in the *NSSC*.[66] Many security professionals consider employee training and education to be a top priority.[67] This can be especially challenging because of the continuously evolving nature of the cybersecurity environment.

Extent of Problems and Response

No in-depth assessment has been made of the degree to which U.S. organizations overall have established effective cybersecurity governance mechanisms. However, many observers believe that governance is an area of substantial weakness,[68] with some variation among different sectors.[69] Problems have been identified at all levels and scales of governance, from failure of leadership in the executive suite and boardroom to inadequate procedures and undertrained personnel. Those weaknesses have been recognized by both DHS and industry, and some initiatives have been developed to address them.[70] Some prominent ventures are discussed later in this report.

PUBLIC KNOWLEDGE AND PERCEPTION

Given widespread publicity about cyberattacks and the repeated revelations of new threats and vulnerabilities, much of the public appears to be aware of and concerned about the possibility of cyberattacks.[71] Although there appears to be little direct evidence on public awareness and preparedness regarding cybersecurity,[72] many experts believe that both home computer users and many organizations —especially small businesses — are not well prepared to take necessary defensive measures. There are several possible reasons for this lack of preparedness, including the following:

- Cybersecurity currently involves a greater level of technical proficiency than many people feel comfortable with.
- Cyberattacks are comparatively easy to hide.[73] Many victims may be unaware of an intrusion unless it results in financial fraud or theft, and that would likely be discovered well after the intrusion occurred.[74]
- Both technology and threats evolve, and user training and education may not keep pace.
- Many organizations underreport cyberattacks and other security incidents, for several reasons, including concerns about negative impacts on public confidence in the organization.[75]
- There are significant economic disincentives for investing in cybersecurity. Most notably, cybersecurity is preventive, not profit making; cyberattacks are comparatively rare; and effects may be distributed — for example, a compromised computer may be used as a means of launching an attack against targets, rather than being a target itself.

In addition, the degree to which cyberattacks pose a serious homeland security risk is a matter of some dispute. While many experts believe that a major cyberattack by terrorists or other adversaries is a substantial risk, others believe that those risks are exaggerated and that the major concern is cybercrime. As with other terrorist incidents, public perception will probably continue to be shaped to a significant degree by the extent of public knowledge about any major attacks that do occur.

Chapter 4

WHAT ARE THE MAJOR MEANS OF LEVERAGE?

The above discussion illustrates the depth of the challenge faced in developing effective cybersecurity. It also shows the diversity, ubiquity, and importance of cyberspace components and demonstrates that cyberspace includes important elements that might not at first glance be considered part of it. Given that diversity and complexity, one approach would be to restrict attention to those components associated with particularly high levels of risk.

Two limits to such an approach are, first, a focus solely on those components known currently to be at high risk could quickly become obsolete. While there are currently many known vulnerabilities which, if addressed, would substantially improve cybersecurity, future or currently undiscovered vulnerabilities may come from unexpected places. Cybercriminals and cyberterrorists would likely seek out new vulnerabilities as current ones are eliminated — writers of "nuisance" viruses have been doing that for several years. In many ways, cybersecurity involves a kind of arms race, with adversaries and defenders each adapting successively to actions by the other. This arms race is likely to continue as long as information technology and cyberspace continue to evolve at current and expected rates.

Second, some would argue that such a focus would simply be an extension of the current fragmented approach, which is largely reactive — as each new vulnerability is discovered, a new fix is developed — and increasingly costly and ineffective. What is needed, they say, is a strategic approach that is more preventive or even preemptive in nature rather than largely reactive and defensive.[76] Some argue that the best approach is to reduce the incentives for catastrophic attack,[77] rather than focusing on preventing all attacks (if experience with cyberspace so far is any indication, this may be impossible or certainly impractical). Such an approach would suggest a focus on (1) limiting damage, and (2) recovery.

To be effective, any preventive approach should probably be broadly applicable to different organizations and systems. The interconnectedness of cyberspace gives it some of the characteristics of a commons — a kind of public resource for which, in the absence of appropriate controls, costs of use by any individual are distributed broadly to the community of users. Classically, using a limited resource — such as pastureland or a fishery — as a commons promotes overuse and degradation of the resource. It pays each individual to maximize his or her use of the resource — to graze as many cattle or catch as many fish as possible — no matter the consequences to the resource as a whole. This effect has been called the "tragedy of the commons."[78] In cyberspace, costs of poor security are often distributed, because compromised systems may be used in attacks on others, with little impact on the compromised system (see above). In addition, however, those costs may be amplified — a naive user may compromise the integrity of an entire network.[79]

There are several options for broadly addressing weaknesses in cybersecurity such as those discussed in the previous section. The following options will be discussed in this section:

- adopting standards and certification,
- promulgating best practices and guidelines,
- using benchmarks and checklists,
- use of auditing,
- improving training and education,
- building security into enterprise architecture,
- using risk management, and
- using metrics.

This discussion is followed by a brief consideration of the role of economic incentives.

STANDARDS

The broad adoption of established standards, or the development and adoption of new ones, could be one way to improve cybersecurity. One widely used definition of standards is "a prescribed set of rules, conditions, or requirements concerning definitions of terms; classification of components; specification of materials, performance, or operations; delineation of procedures; or measurement of quantity and quality in describing materials, products, systems, services, or practices."[80] As this rather eclectic definition illustrates, there are many different kinds of standards.[81] They may be classified according to purpose — e.g., product, process, testing, or interface standards. They can also be classified according to their focus — commonly, a distinction is made between performance standards, which focus on function, and design standards, which specify features, dimensions, or other such characteristics. A third classification is based on how standards are developed and implemented. They may be developed through consensus or some other process. They may be implemented voluntarily, or they may also be imposed, for example by law, and therefore mandatory. Voluntary consensus standards are common, and federal law encourages their use by federal agencies, including DHS.[82] Standards may also be open or proprietary, but different observers define "open standard" somewhat differently.[83] Some form of open standards is the approach used typically by major standards organizations.

Which kinds of standards to adopt will very much depend on the goals identified and the characteristics of specific elements. In general, design standards or detailed regulation usually cannot evolve readily in parallel to an evolving technology. Given the rapid evolution of information technology, there appears to be agreement that their use should be avoided for elements that are not yet mature if appropriate results can be obtained through more flexible approaches, such as performance standards or best practices.

Several organizations are involved in the development of cybersecurity standards. NIST performs a wide array of standards-related activities, including promoting the global use of U.S. standards, providing information and technical support to industry and others, coordinating the development of national voluntary product standards, accrediting testing laboratories, and developing standards for use by federal agencies where no acceptable industry standards exist.[84] The American National Standards Institute (ANSI) is a private, nonprofit organization that administers and coordinates the U.S. voluntary private-sector standardization system.[85] ANSI and NIST coordinate activities through a memorandum of understanding.[86] Among ANSI's activities related to cybersecurity are its Information Systems Conference Committee, which provides a forum for communication among IT standards developers, and the Information Infrastructure Standards Panel, which identifies standards critical for global information infrastructure. While ANSI also has established a homeland security standards panel, cybersecurity is not among the panel's areas of focus. NIST activities include the Process Control Security Requirements Forum, which is developing security requirements for industrial process control systems.[87]

Among other U.S. organizations engaged in standards activities related to cybersecurity are the InterNational Committee for Information Technology Standards[88] and the Institute of Electrical and Electronic Engineers.[89] The Trusted Computing Group[90] is a group of IT manufacturers, vendors, and others formed in April 2003 to develop open industry hardware and software standards for trusted computing, an important element of cybersecurity. The Internet Engineering Task Force (ITEF)[91] is an international group of experts and others involved in the development and operation of the Internet; participation is open to any interested person.

The International Organization for Standardization (ISO),[92] a nonprofit network of national standards organizations from various countries, is the major international standards developer. The International Electrotechnical Commission (IEC)[93] develops standards relating to electronic technologies. Together they have established a Joint Technical Committee on Information Technology (JTC1),[94] with a subcommittee on security techniques (JTC1 SC27)[95] that develops generic standards relating to IT security.

Current Standards

Several sets of standards have been developed for use in cybersecurity. Three of the most widely cited are the Common Criteria for Information Technology Security Evaluation (usually called the Common Criteria, abbreviated CC); ISO/IEC 17799, an internationally recognized information security standard; and the Federal Information Processing Standards (FIPS), which were developed by NIST for use by federal systems. These are each discussed below. A wide range of international standards also exist for specific security techniques, such as encryption, authentication, nonrepudiation, and time stamping.[96]

Product Evaluation

The Common Criteria consist of a set of evaluation criteria for the security of information technology that was developed by U.S., Canadian, and some European government agencies. It resulted from a recognition of the need to harmonize separate evaluation criteria that had been developed by different countries.[97] It was also adopted as an international technical standard (ISO/IEC 15408) in 1999. The CC provides a framework for the development of standard sets of requirements, called profiles, to meet specific needs of consumers and developers of information technology products, depending on the assurance levels that they require.[98] A set of *protection profiles* may be developed for different kinds of products (such as a firewall) or general applications (such as electronic fund transfers) that may be evaluated.[99] The profiles lay out security objectives and requirements. For example, a profile developed for Department of Defense firewalls describes the security environment to which the profile applies, threats to be addressed, security objectives, functional and assurance requirements to meet those objectives, and the rationale for how the requirements meet the objectives and how the objectives address the threats.[100] Once developed, a profile may be evaluated by an accredited, independent laboratory. More than 40 profiles have been developed for a range of products and systems, and most have received evaluations.

For a specific application,[101] a set of security requirements and specifications[102] is developed, usually conforming to one or more relevant protection profiles if available. The application is then evaluated to determine if it meets those requirements and specifications, and if so, it may be certified for use in the specified environment. Products may be evaluated to any of seven hierarchical *evaluation assurance levels* (EALs), which reflect successively higher levels of security. Both software and hardware products have been certified under the CC. They include operating systems, databases, firewalls, computer chips, smartcards, and routers, among others.[103] More than 100 applications have been evaluated at EAL1 to EAL4+.[104]

Although the CC are often referred to as standards, there are aspects of them that are not easily characterized as standards, at least according to some observers. The notion of criteria is broader than that of standards because it generally includes things, such as statements on how a system should be designed and operated, that cannot be directly assessed by examining the product.[105] Also, protection profiles are not written into the CC but are developed and updated as needed.

Code of Practice

Several standards have been developed relating to overall information security practices. They might be used in conjunction with other guides such as the CC as elements of an overall framework for cybersecurity. There appears to be at least some agreement that a good security management standard should cover all important security issues; be comprehensive and up-to-date; be clear, unambiguous, and easy to understand and use; be practical and achievable; be scalable to any organization; and provide a basis for measurement of performance.[106]

The most widely recognized code-of-practice standards are ISO/IEC 13335 and ISO/IEC 17779. The first provides broad guidelines for managing IT security

(GMITS) in the context of an organization's overall management, and stresses challenges posed by the global nature of cyberspace. It addresses universal security concepts, management and planning, risk assessment, merits of alternative solutions, and external communications. It focuses on high-level concepts and general requirements and techniques, rather than specific controls. It describes IT security management as including a determination of objectives, strategies, policies and organizational requirements; managing risks; planning implementation of adequate safeguards and follow-up programs for monitoring, reviewing, and maintaining security services; and developing a security-awareness program and plans for incident-handling. It was released in parts, including five technical reports, from 1996 to 2001. A revision was begun in 2000.[107]

ISO/IEC 17799 is described by JCT1 SC27 as giving "recommendations for information security management for use by those who are responsible for initiating, implementing or maintaining security in their organization. It is intended to provide a common basis for developing organizational security standards and effective security management practice and to provide confidence in inter-organizational dealings."[108] Topics covered include

- organizational policy and infrastructure;
- asset classification and control;
- personnel, physical, and environmental security;
- communications and operations management;
- access control;
- systems development and maintenance;
- business continuity; and
- compliance.[109]

ISO/IEC 17799 is more widely recognized internationally than any other cybersecurity management standard.[110] It is related to ISO/IEC 13335 in that "17799 focuses on issues to be considered for information security management and…13335 addresses how to achieve [it]."[111] The standard was issued in 2000, and revision began in 2001. It is based on and virtually identical to the 1999 update of the British Standard in Information Security, BS 7799 (Part 1), which was initially published in 1995.[112]

While called a standard, ISO/IEC 17799 has been described as more similar to a set of guidelines, in that it is not written in such a way that conformance can be certified.[113] The standard contains 127 major controls and thousands of bits of guidance, but they are not presented as imperatives.[114] Thus, organizations may adapt the standard to their needs, modifying the application of some sections to fit their management structure, or discarding sections that do not apply.[115] This flexibility has been both praised and criticized. On the one hand, it means that organizations can use the standard without compromising other key business requirements. On the other hand, it makes conformance more difficult to assess.[116]

While ISO/IEC 17799 does not itself include a certification scheme, some countries have developed such schemes. Perhaps most notable is BS 7799 Part 2, developed and used in Great Britain and also available in other countries, including the United States.[117] This standard specifies requirements and controls for an organization's information security management system (ISMS) in ways that can be assessed by an accredited certification body. It has been described as consisting of requirements for an ISMS plus ISO/IEC 17799 controls[118] "in imperative format."[119] The most recent version of BS 7799 Part 2 was published in 2002. There does not appear to be any equivalent under development for ISO/IEC 17799 itself.

The Information Security Forum (ISF) has developed a code of practice, *The Standard of Good Practice for Information Security*.[120] ISF updates the standard every two years. It was first released in 1996, with the most recent version released in March 2003. It is based on the experience and expertise of ISF members and staff, other standards such as ISO/IEC 17799, and the results of ISF surveys. Topics covered include security management, critical business applications, computer installations, networks, and systems development.

That set of topics appears somewhat more limited in scope than the set covered by ISO/IEC 17799, but a direct comparison was not possible for this report. Each topic is organized into several areas (30 altogether), which in turn contain several sections (132 altogether). Each section contains a principle, an objective, and several specific actions or controls. The IFS standard is publicly available without charge, unlike ISO/IEC 17799.[121] IFS provides members with a survey instrument they can use to compare their performance against the IFS standard and other benchmarks, but the organization does not appear to provide certification.

The IT Governance Institute (ITGI)[122] has developed *Control Objectives for Information and related Technology (COBIT),* a set of recommended practices in information technology governance, control, and assurance developed through a consensus process involving experts. First released in 1996, the third edition was published in 2000. It provides a framework for IT governance, including metrics and other management tools as well as controls. ITGI does not describe COBIT as a standard but alternatively as a "framework for IT governance"[123] and a "generally accepted best practice."[124] Nevertheless, it is similar enough in both structure and method of development to the standards described above that it arguably should be considered a code-of-practice standard. Rather than specifically focusing on cybersecurity, it addresses security in the context of overall IT governance. Security is considered one of three sets of requirements, the other two being quality and fiduciary. COBIT is organized hierarchically into four domains, which are broad categories of activity such as planning, implementation, and monitoring; 34 processes; and specific activities or objectives under each process.[125] There is no certification program for COBIT, but audit and self-assessment guidelines are available. The framework has been criticized as being difficult to scale to small or medium-sized enterprises, but ITGI has developed a version aimed at such organizations.[126]

Federal Standards

NIST is responsible under federal law[127] for developing standards and guidelines for cybersecurity for federal information systems, except national security systems, which fall under the responsibility of the Committee on National Security Systems (CNSS) and the agencies that operate the systems.[128] The Federal Information Processing Standards (FIPS) are standards developed by NIST for requirements for federal systems not covered by available voluntary industry standards.[129] Some FIPS are mandatory for federal agencies, while others are not. FISMA requires NIST to "develop standards and guidelines, including minimum requirements, for providing adequate information security for all agency operations and assets," except for national security systems.[130] None of the FIPS publications to date specifically address governance issues.

FIPS are developed with rule-making procedures similar to those established by the Administrative Procedure Act.[131] Some FIPS are adopted by private sector entities. For example, the Data Encryption Standard (DES — FIPS 46), introduced in 1977, provides a method for cryptographic protection of information. It was widely adopted by industry, for example in the financial services sector. The newer, stronger Advanced Encryption Standard (AES — FIPS 197), adopted in 2001, is now replacing DES as applications are developed.

In its series of special publications on computer security,[132] NIST has published a set of generally accepted system security principles and practices[133] (sometimes called GAPP), discussed earlier in this report, that are similar in scope to ISO/IEC 17799, and the two are sometimes considered to be competing standards. No general certification scheme exists for this set of practices. There are also several other NIST publications on various aspects of cybersecurity, such as capital planning, system development, security awareness and training, and so forth.[134]

NSA has established an Information Assurance Technical Framework Forum (IATFF)[135] to develop a framework for solutions to information assurance problems encountered by federal agencies and industry. A framework document[136] available through the forum provides technical guidance for protecting information and information infrastructure using NSA's defense-in-depth strategy.

Strengths and Weaknesses of Standards

The widespread use of well established and well-designed cybersecurity standards would have potential benefits. Such standards would provide a common language and criteria for determining how well organizations are adhering to recognized security needs and requirements. In addition, as the use of the standards increased, the overall level of security would arguably rise as well. Also, the standards would presumably provide a common baseline from which continuous improvement in cybersecurity could be implemented through the evolution of the standards.

However, the use of standards in cybersecurity has also been criticized by some. Some common criticisms are described below.

They Are Not Sufficiently Flexible and Cannot Track Changes in the Technology

International standards are often updated on a three- to five-year cycle. Given the rate of evolution of cyberspace, some observers have complained that standards become outdated too quickly to be useful for cybersecurity. Proponents counter that properly developed standards are in fact sufficiently flexible that they can accommodate the technological improvements that are likely to occur between revisions. International standards such as ISO 17799 are often revised on a three- to five-year cycle. Both COBIT and the ISF standard are updated on a two-year cycle. The Common Criteria Development Board is charged with issuing updates and corrections to the CC.[137]

They Can Be Expensive to Conform to

If certification is available, as with BS 7799 Part 2, the process of becoming certified may be expensive, especially for smaller enterprises. Even without certification, organizations adopting standards may find they need to significantly alter business practices, possibly at considerable expense and sometimes in ways that are not in keeping with the optimum business model for the particular enterprise. Proponents counter that, while return on investment may be difficult to measure directly, the process of coming into compliance can help organizations identify and correct serious cybersecurity deficiencies, and protect them from large expenditures to recover from a success attack or from loss of reputation that can be very difficult to regain.

They Are too Much Like Regulation

If adherence to a particular set of standards becomes expected, then certification bodies might take on some of the characteristics of regulators, with the attendant benefits and disadvantages. Proponents may counter that such need not be the case, especially if the standards and certification are well designed, there are sufficient alternative paths to certification to avoid the development of effective monopolies, and compliance is voluntary, as it is with most standards.

The mixed success of the Common Criteria illustrates some of these reported pitfalls. These include a lack of flexibility, despite attempts to build flexibility into the CC; the inability to keep pace with evolving technology; and cost and time required for certification.[138]

Measuring Success May Be Difficult for Code-of-Practice Standards

"High level" code-of-practice standards such as ISO/IEC 17799 have been criticized for not being specific enough to provide sufficient guidance or a sufficient common basis for measuring and comparing practices among different organizations. At the same time, BS 7799 Part 2 has been criticized for being too much of a checklist and insufficiently adaptable to different kinds of enterprises. Proponents counter that such critics misunderstand the application of the standards — that comparable metrics can be developed and that certification can readily be adapted to the requirements of a particular enterprise. CC, COBIT and other standards have been criticized for being difficult to scale, especially to the needs of smaller organizations that may not have a primary IT focus. Attempts have been made to compensate for this problem. For example, ITGI has developed a form of COBIT specifically designed for smaller enterprises. Despite such concerns, the advantages of code-of practice and other cybersecurity standards appear to be sufficient that their use is increasing (see below).

The Development Process May Be Cumbersome

Some of the criticisms associated with standards result from the particular methods by which most standards are developed. For example, the ANSI process includes "consensus on a proposed standard by a group or 'consensus body' that includes representatives from materially affected and interested parties; broad-based public review and comment on draft standards; consideration of and response to comments submitted by voting members of the relevant consensus body and by public review commenters; incorporation of approved changes into a draft standard; and right to appeal by any participant that believes that due process principles were not sufficiently respected during the standards development in accordance with the ANSI-accredited procedures of the standards developer."[139] The designated "consensus body" is required to be balanced with regard to different interests. Consensus does not require unanimity but does require "substantial agreement...by directly and materially affected interests...[and] that all views and objections be considered, and that an effort be made toward their resolution."[140] This process, which may require several meetings, ensures that the interests of all involved parties are taken into account, but it can be slow and may require compromises that can lead to more complex standards.

In contrast, the Internet Engineering Task Force (IETF) develops standards through a process that is performed largely online. Interested parties form a working group to identify the scope of the standard and begin developing it. Participation in the working group is completely open to anyone interested, but there is no active attempt to guarantee a balance among different interests. Drafts of the standard are posted online and comments incorporated. Once the group reaches a "rough consensus," defined as agreement by a "very large majority" of the working group,[141] the draft is sent to the Internet Engineering Steering Group (IESG) for independent review by experts. After successfully passing review, the draft may become a standard through some additional steps. According to some observers, the use of a fully open, online process, rough consensus, and independent review results in "cleaner" standards and a more rapid process than the more traditional approach taken by most standards bodies.

CERTIFICATION

Certification usually refers to a formal approval by some entity, such as a laboratory, that a product, process, or person meets a specified set of criteria. For example, an electrical product may be certified as meeting safety standards. A physician may be certified as meeting a particular level of competency in an area of specialization. The certifying entity may be *accredited* by a recognized authority such as a government agency or professional association. Accreditation may also refer to the approval of a certified product for use in a particular system[142] or it may refer to the authorization to use a particular information system and accept the attendant risks.[143]

Certification processes exist for both product evaluation and code of practice standards. For example, products can be certified under the CC, as discussed above. Other product evaluation certifications have also been developed. For instance, the Technology Group for The Financial Services Roundtable (BITS) runs a security certification program for products used by the financial services industry.[144] The criteria used follow the general scheme laid out in the CC. For code of practice, certification is available in many countries, including the United States, under BS7799 Part 2. The number of those certifications has been increasing substantially, especially in Asia,[145] with more than 800 organizations certified worldwide, although only a few in the United States.[146]

Professional certification is also available from some organizations. For example, the Information Systems Audit and Control Association (ISACA)[147] offers certification for information security auditors and managers, and the International Information Systems Security Certification Consortium[148] offers certification for information security professionals. Such certifications usually require several years of relevant professional experience, successful completion of an examination process, adherence to a code of conduct, and continuing education in the field.

Strengths and Weaknesses of Certification

Certification can be an important component of any attempt to adhere to a set of established standards. That is because it provides a means of independent verification that criteria set by the standards have been met. Many of the criticisms of standards discussed above, and counters to them, can be applied to certification as well.

The strengths and weaknesses of certification can be illustrated by ISO/IEC 17799 and the CC. If a certification were available for ISO/IEC 17799, companies that claim to have adopted it could demonstrate that they have been assessed by an independent, accredited body as conforming to its requirements. However, they would not be free to adapt the standards however they wished to their particular operating situations and needs. A product certified under the CC can be used with confidence in the kinds of environments to which the certification applies. However, the certification process is expensive and time-consuming, increasing the costs of the products and potentially impeding the adoption of newer technologies.

There also does not appear to have been any systematic assessment of the effectiveness of certification under standards such as BS7799 Part 2 with respect to improving cybersecurity. That may be in part because the certification has been available for only a few years. There are at least two ways that success could be measured and that different standards and methods of compliance could be compared.[149] First, the incidence of security problems (including but not limited to attacks) would be expected to be lower for organizations using the most effective standard and compliance method. That measure may be hard to use as long as organizations are reluctant to reveal security breaches or other problems, as has been reported.[150] Another, more indirect metric would be the relative success of different certifications. Presumably, an organization that finds a particular certification to be effective would be more likely to renew it — or to purchase additional products certified under it — than switch to another or discontinue use. However, other factors, such as cost, can also influence the relative success of different certification regimes.

BEST PRACTICES

Best practices often refers to strategies, policies, procedures, and other action related elements of cybersecurity that are generally accepted as being the most successful or cost-effective. Such practices can be identified for virtually any of the elements of a cybersecurity framework, from goals to specific procedures or specifications.

Unfortunately, there does not appear to be any overall agreement on what constitutes a best practice. The term implies that the practice in question has been assessed as being superior to all others, but the basis of such assessments, if provided, usually appears, at best, to be a consensus among experts, rather than a rigorous empirical comparison of alternatives. In fact, it is not uncommon in the literature for a set of "best practices" to be asserted with no description of what criteria were used to identify them as best. Given the vagueness associated with the use of this term, it might be more appropriate to refer instead to *commonly accepted* or *generally accepted practices*, at least where there is evidence to that effect.[151]

What is called a set of best practices can vary greatly in content and method of development. At one extreme are standards developed through a well-established methodology, such as the code of practices contained in ISO/IEC

17799 or COBIT. At the other extreme, a set of "best practices" might simply be recommendations from one person published in a newsletter article. Best practices may be developed specifically for one sector or industry. For example, the Network Reliability and Interoperability Council (NRIC) has developed a set of more than 150 cybersecurity best practices for the communications industry.[152] Most of these are fairly general, such as "disable unnecessary services" but some are much more specific. However, they are intended to address classes of problems rather than providing "[d]etailed fixes to specific problems…." NRIC used an "industry consensus" approach to develop them, stressing that a practice is included only after "sufficient rigor and deliberation" including "[d]iscussions among experts and stakeholders" about whether the practice is implemented widely enough, its effectiveness and feasibility, the risk associated with failing to implement it, and alternatives. NRIC proposes that these practices be used as recommendations and not as requirements and that they be adapted to the individual needs of the organization using them. In another example, the ASP Industry Consortium produced a set of white papers, prepared by the consortium's security subcommittee, that include about 25 best practices for network and platform security.[153] The practices described are fairly general, such as "use remote access sparingly." The methodology by which they were developed is not described.

Another group of best practices with relevance to cybersecurity is known as capability maturity models (CMM). Essentially, these are practices, arranged along a hierarchy of maturity levels, designed to help organizations identify the level at which they operate processes for the development of software and other products and to improve those processes by successively improving to higher levels of maturity.[154] The system has been developed as a joint public-private partnership initiated by the Department of Defense in the 1980s. One example is "cleanroom software engineering" — procedures based on mathematical verification of designs and statistical testing of systems that are designed to produce highly reliable software that has a minimum of errors and vulnerabilities. For applications where security considerations are a priority, techniques have been developed to engineer systems to the appropriate level of security corresponding to the specific needs for the application. Such systems are designed with carefully specified requirements and are thoroughly reviewed and tested before implementation.[155]

Best practices would not necessarily be associated with a certification or audit process, so it can be difficult to determine if an organization is in fact conforming to them effectively. However, they generally provide a degree of flexibility and adaptability that may not be available with more formal approaches. Furthermore, they may be easier to update in response to the rapid evolution of technology, cyberspace, and the threat environment.

GUIDELINES

Guidelines may be thought of as general recommendations relating to elements of cybersecurity. They are not necessarily associated with any particular methodology or criteria, in contrast to standards and (at least in theory) best practices, other than the authority of those making the recommendations. One commonly cited set of guidelines is the *Guidelines for the Security of Information Systems and Networks* of The Organization for Economic Cooperation and Development, first adopted in 1992 and most recently revised in 2002. The nine basic principles contained in the guidelines are intended to provide a foundation for the development of a "culture of security." The principles focus on the importance of awareness of and responsibility for security, the importance of timely responsiveness to security incidents, the role of ethical considerations and democratic values, the need for risk assessments, security as an essential design element for information systems, the need for comprehensive security management, and the importance of continual review and reassessment. Many of these principles are also reflected in other documents, including ISO/IEC 17799.

Generally Accepted Information Security Principles. GAISP is an attempt to draw together a hierarchical set of principles that have been reviewed by experts in information security and that meet specified criteria. The project was initiated by the Information Systems Security Association, an international, nonprofit association of information security professionals. GAISP consists of "principles, standards, conventions, and mechanisms that information security practitioners should employ, that information processing products should provide, and that information owners and organizational governance should acknowledge to ensure the security of information and information systems."[156] It is intended to provide a basis for self regulation for the profession, analogous to the Generally Accepted Accounting Principles (GAAP) used by Certified Public Accountants.[157] The hierarchical approach aims to provide guidance that can be applied at various levels within an organization, from executive governance to daily management of security risks.

Basel Principles. The financial services sector has been among the leaders in developing and implementing components of a cybersecurity framework. The Basel Committee on Banking Supervision has released a set of guidelines called *Risk Management Principles for Electronic Banking*.[158] While seven of the fourteen principles and practices described in the document relate to security controls,[159] the Basel principles are particularly notable for the degree to which they stress the importance of institutional leadership and the management of legal and reputational risk in the context of cybersecurity. For example, the first three principles place responsibility for active oversight of cybersecurity management directly on boards of directors and senior management. The principles relating to legal and reputational risk focus on information disclosure, protection of customer data, including privacy, and continuity of service.

The difference between guidelines and best practices is not perhaps as distinct as the difference between either of those and standards. While guidelines may provide even greater flexibility and adaptability than best practices, their general lack of specificity may make effective implementation more challenging. As with best practices, guidelines would not necessarily be associated with a certification or audit process, so it might be difficult to determine if an organization is in fact conforming to them effectively.

BENCHMARKS AND CHECKLISTS

Fundamentally, a *benchmark* is simply a reference point against which performance is measured. It might be used as a goal, or it might be considered a level of minimum acceptable performance. The latter might also be called a *baseline*. With respect to computers, a benchmark often refers to a test used to compare one or more aspects of performance of a system (such as processing speed) with other systems or with a specified level of function.

With respect to cybersecurity, the terms *benchmarks* and *checklists* are more often used to denote sets of security configurations and settings that are recommended to achieve a specified level of security. One well-known set provides minimum security configurations for the Microsoft Windows 2000 operating system. Developed through a consensus process involving federal agencies and private organizations,[160] it was released by the Center for Internet Security (CIS) in 2002.[161] Security configuration benchmarks have also been developed for other operating systems, application software, and some hardware.[162] NIST has developed a program to devise security checklists for software and hardware used by federal agencies.[163] The Defense Information Systems Agency (DISA)[164] and NSA also produce configuration guidance documents.

Producing an effective set of code-of-practice benchmarks is arguably more difficult than producing technical configuration guidance. One example of a set of code-of-practice benchmarks was developed by the Human Firewall Council, a consortium of information security professionals. Called the Security Management Index, it is now managed by ISSA.[165] Based on ISO 17799, it permits organizations to perform self-assessments, via completion of a survey, to determine how well they conform to the objectives in the standard in comparison to other organizations that have participated.

Benchmarks and checklists can be an important element of a cybersecurity framework but are by their nature very specific and limited in scope. Also, some confusion may result from the occasional use of the term as a synonym for standards.

AUDITING

Auditing is often thought of as a formal examination of financial or accounting records, but it is also used in a broader sense, such as to denote independent examination of an organization's adherence to established controls, policies, or legal requirements.[166] An organization may undergo, for example, a security audit of its information systems. That may involve an examination of hardware, software, procedures, configurations, environment, and user practices. An audit may be performed by the organization itself, or it may be performed by an independent auditor, usually a firm that specializes in accounting and auditing. Audits usually follow a set of established practices and procedures, such as the Statement on Auditing Standards No. 70 (known as SAS-70) issued by the Auditing Standards Board of the American Institute of Certified Public Accountants (AICPA).[167] Information security audit guides have also been developed for government agencies.[168] An audit usually involves testing of controls and results in a report that includes the opinion of the auditor about the adequacy of the controls examined, with recommendations for improvements. It does not result in a certification of conformance to a standard. However, auditors may be expected to conform to established standards in the conduct of an audit.[169]

Auditing methods and requirements are most well developed with respect to financial and accounting processes. As a result, some audits might tend to underemphasize aspects of cybersecurity that are not related to those processes. The results of audits might also vary significantly among different auditors. The Sarbanes-Oxley Act of 2002 (P.L. 107-204) requires audits of financial controls, including information security controls, for publicly traded companies.

Training and Education

If, as some observers believe, people are the most important element of effective cybersecurity, then training and education should be an important means of leverage to improve cybersecurity. Inadequate cybersecurity practices by users, IT personnel, and even corporate leadership have been widely cited as a major vulnerability.[170] The *NSSC* lists national cyberspace security awareness and training as one of its top five priorities. Elements include a comprehensive national awareness program and support for training, education, and professional certification.[171] The National Cyber Security Alliance (NCSA) has been established as a public-private partnership of government agencies, corporations, and nongovernmental organizations to promote cybersecurity education and awareness.[172]

Many factors can influence the effectiveness of training and education to enhance cybersecurity. For example, programs and materials vary in quality, and poorly designed program is unlikely to provide significant improvements in cybersecurity. In addition, training may not be able to compensate sufficiently for a poor system design.

Enterprise Architecture

Effective cybersecurity needs to focus not only on the individual elements of an organization's information technology but also how they interact. The term *enterprise architecture* (EA) has become increasingly used to refer to the components of an organization and how they work together to achieve the organization's objectives. Specific definitions and usage vary. EA is often used specifically to refer to the information technology component of the architecture, and especially to the interoperability of those components. It is also used to denote a blueprint of an organization's business operations and the technology required to perform those operations.[173] The federal government is developing a "business-driven" EA to improve interoperability and services.[174]

An organization can characterize its EA to assist in planning and development of its information technology. Such a characterization can provide an opportunity to make security an integral part of EA. This component of EA is sometimes called the security architecture.[175] However, even the initial characterization of an organization's EA can be time-consuming and expensive, and the costs of reengineering to build in security may be prohibitive for many organizations. In addition, the need to build a business case to justify IT investments, which is often considered important to the EA approach, may create barriers to improving security, given the traditional difficulties of demonstrating a financial return on investments in security.

Risk Management

The approach embodied in defense-in-depth recognizes that security cannot be perfect but rather reduces the risk and impact of a successful attack or other breach. Such reduction can be captured through risk management, which involves identifying, controlling, and mitigating threats, vulnerabilities, and the impacts of security breaches. The steps in effective risk management include assessment of risk, steps to mitigate them, and continuous evaluation and adjustment. The approach often involves cost-benefit analysis to help determine optimal mitigation steps. Mitigation may involve accepting the risk as a cost of business; avoiding risk associated with a particular activity, for example by not engaging in it; limiting the risk through effective use of controls; and transferring the risk, for example through insurance.[176] Some insurance companies have begun to offer cybersecurity policies, under which companies can transfer some of their risks in the event of a successful attack. Carriers may require clients to implement specified security practices to qualify for insurance. However, in the absence of reliable actuarial data about the risks and costs of cyberattacks, it may be difficult for carriers to set appropriate insurance rates.

To be effective, risk management requires accurate risk assessment. However, many cybersecurity risks may be difficult to assess, for reasons discussed earlier. In addition, a risk management approach may lead an organization to accept risks for which the potential impacts of security events are low, regardless of external impacts.[177] Thus, risk management is not likely to sufficiently address cybersecurity problems associated with the commons properties of cyberspace discussed earlier in this report.

Metrics

Whatever approaches are used to improve cybersecurity, measuring their success would appear to be essential to determining how effective they are and to making improvements. However, fundamental problems exist with measuring success in security. Seemingly, the most appropriate measure is the number of successful attacks, but in fact, attacks — especially the kind of major attack for which effective defense is critical — may be comparatively uncommon, so that absence of a successful attack may not indicate effective security.[178] In addition, attackers often take steps to avoid detection, so an absence of detected attacks may in fact be a measure of poor rather than good security. This conceptual problem might be addressed through the use of proxy measures, such as how well technology, policy, and activities conform to certain accepted benchmarks, as well as the use of proficiency testing, such as blind "red team" attacks or other penetration testing.

Not only is it difficult to identify appropriate metrics for cybersecurity, there are also risks of distortions that may be associated with any particular metric. Virtually any given metric will measure only one or a limited number of aspects of a goal. If, however, the limitations of the metric are not understood, attempts to use it to optimize security can lead to distortions, as the above example illustrates. This appears to be a general concern.[179] However, some argue that using even distorted metrics can be beneficial if the process of measuring them focuses attention on problems or deficiencies and leads to correction.

Metrics relating to the effects of security events are called *impact metrics.* Those relating to the delivery of security services are called *effectiveness* or *efficiency metrics;* and those relating to the execution of security policies are called *implementation metrics.* NIST has published guidelines on such metrics, to assist agencies in complying with federal requirements.[180] The document does not urge the adoption of any specific set of metrics, although it does provide examples. Instead, it recommends that the metrics chosen use data that can be realistically obtained, that measure existing, stable processes, and that facilitate the improvement of security implementation. The kinds of metrics that can be effectively gathered will depend on the level of maturity of the security program. Programs at low levels of maturity will of necessity be limited to using implementation metrics. Impact metrics can be effective for organizations that have mature security programs, with fully integrated procedures and controls.[181]

Economic Incentives

Implementation of cybersecurity measures may involve substantial costs and is therefore sensitive to market forces and other economic factors. If sufficient economic incentives exist for improving cybersecurity, then organizations are likely to make the investments needed in the absence of government regulation or other drivers. One concern often raised is that economic incentives are often insufficient, and that in fact, significant counterincentives exist.

The perceived inadequacy of incentives for cybersecurity can be seen as a form of *market failure* — a kind of economic inefficiency.[182] There are several lines of evidence supporting this view. For example, it can be difficult for law enforcement officials to arrest and prosecute hackers if companies are unwilling to provide information on cyberattacks, yet a company risks suffering significant reputation costs if that information leads customers to conclude that the company's information systems are not sufficiently secure. In addition, investments in cybersecurity cannot easily be analyzed in terms of return on investment, since they do not contribute to income in a measurable way.[183] Therefore, companies may be reluctant to make the necessary investments. Also, impacts of compromised systems may reach far beyond the system where the compromise occurred[184] — the interconnectedness of cyberspace has made it to a significant extent a commons, with associated economic externalities.

The widespread adoption of the kinds of leverage to improve cybersecurity discussed above may be doubtful without changes in the current incentive structure. Such changes could arise from several sources. Among them are increases in public demand for cybersecurity, changes in expected behavior within a sector regarding investment in cybersecurity,[185] public-private partnerships, and regulation or other action by governments. While not all such factors are themselves economic in nature, they can clearly affect the economic incentive structure. For example, a company that does not respond to expectations from its peers for improved cybersecurity may suffer a significant reputation cost. Similarly, a company that is found to violate government requirements may suffer both reputation costs and direct punitive action or may be held financially liable for damages.

Chapter 5

What Roles should Government and the Private Sector Play?

The above discussion shows that (1) there is currently no unified national framework for improving cybersecurity, (2) there are several areas of weakness where such a framework could be useful in generating improvements, and (3) several means of leverage exist that could be used in the development and implementation of such a framework. Questions remain, however, about whether additional federal efforts would be needed or desirable. According to the *NSSC*,

> "a government role in cybersecurity is warranted in [nongovernmental] cases where high transaction costs or legal barriers lead to significant coordination problems; cases in which governments operate in the absence of private sector forces; resolution of incentive problems that lead to under provisioning of critical shared resources; and raising awareness."[186]

Are market forces, along with current government and private-sector policies and practices, sufficient to put in place the necessary components? If not, will additional voluntary efforts be sufficient, or is further government action required? This section discusses whether current efforts are adequate and what policy options exist for further action.

Current Efforts

While many observers argue that cybersecurity efforts remain inadequate overall, substantial evidence of improvements can be found. They range from steady increases in the number of organizations adopting cybersecurity standards of practice (see discussion above) to efforts to increase public awareness about cybersecurity to new federal and state requirements for government and private-sector information systems. The legal framework for cybersecurity continues to evolve, with new federal and state laws being implemented, and new public-private partnerships have been developed.

Laws and Regulations

According to the *NSSC,*

> It is the policy of the United States to prevent or minimize disruptions to critical information infrastructures and thereby protect the people, the economy, the essential human and government services, and the national security of the United States. Disruptions that do occur should be infrequent, of minimal duration and manageable and cause the least damage possible. The policy requires a continuous effort to secure information systems for critical infrastructure and includes voluntary public-private partnerships involving corporate and nongovernmental organizations.[187]

However, current federal law and regulation is generally much narrower in scope, applying to computer systems operated by or on behalf of the federal government.[188] The requirements and governance mechanisms differ depending on whether or not a system is designated as a national security system. In general, however, the Federal Information Security Management Act of 2002 (FISMA, title III of P.L 107-347, the E-Government Act of 2002) requires agencies to develop policies and standards to provide for the integrity, confidentiality, and availability of information. As required by the act, NIST has developed a broad range of standards and guidelines,[189] and the Office of Management and Budget (OMB) reports annually to Congress on agency compliance with IT security requirements. In its most recent annual report, OMB reported substantial improvements overall,[190] although the Government Accountability Office (GAO) noted wide variability in agency compliance and "significant weaknesses…that put critical operations and assets at risk."[191] Among the areas of weakness cited was the program management framework for security.

Despite remaining weaknesses and concerns, these federal programs and requirements can be important not only directly, by improving federal cybersecurity, but also by providing information, opportunities, and incentives for improving cybersecurity in the private sector. NIST's FISMA standards and guidelines are publicly available, as are some information assurance documents produced by the National Security Agency.[192] Even though they apply only to government agencies and contractors, federal cybersecurity requirements can also potentially stimulate a market for more secure products. For example, if a company's product must meet certain security specifications for federal agencies, it may be more cost-effective for the company to make those specifications available in general rather than customizing the product for the federal government. Nevertheless, private-sector organizations are not required to implement FISMA, and its impact on nongovernmental cybersecurity does not appear to be well-characterized.

Some federal laws do place security requirements on certain classes of private sector information and controls.[193] These include protections for personal information for customers of financial institutions (Gramm-Leach-Bliley Act of 1999, P.L. 106-102), health information that is held by health-sector entities and that is identifiable with respect to a person (Health Insurance Portability and Accountability Act of 1996, P.L. 104-191), and audits of financial controls, which has been interpreted as including information security, of publicly registered companies (Sarbanes-Oxley Act of 2002, P.L. 107-204).[194] These are obviously limited domains of influence in cyberspace, but they appear to have resulted in significant response in the private sector[195] and may have influence beyond their immediate domains of applicability.[196] The Sarbanes-Oxley Act in particular impacts corporate governance with respect to cybersecurity by specifying that corporate management is responsible for establishing and maintaining adequate internal controls.[197] Nevertheless, none of these laws specifically address the question of a framework for cybersecurity. Their major influence on the development of such a framework may be the regulatory incentives that they provide for corporate management to address cybersecurity issues.

The Homeland Security Act of 2002 (HSA) gives DHS some authority and resources relating to cybersecurity. The National Cybersecurity Division (NCSD) was established in June 2003 within the Directorate for Information Assurance and Infrastructure Protection (IAIP) of the department. According to DHS, the division's mission includes the following: "(1) identifying, analyzing, and reducing cyber threats and vulnerabilities; (2) disseminating cyber threat warning information; (3) coordinating cyber incident response; and, (4) providing technical assistance in continuity of operations and recovery from cyber incidents."[198] NCSD has created a computer emergency response team, US-CERT,[199] in cooperation with Carnegie- Mellon University, to coordinate cybersecurity efforts, and established a new alert system. It has also engaged in efforts to facilitate public-private cybersecurity partnerships, notably by sponsoring the National Cybersecurity Summit to that end in December 2003, and follow-up efforts. DHS also sponsors cybersecurity research and development within its Science and Technology Directorate.

State laws can also have impacts both within and beyond the states that enact them. For example, the California Database Protection Act (CA S.B.1386), which went into effect July 1, 2003, requires any government or private entity doing business in California to reveal to affected residents of the state any security breach that results in unauthorized acquisition of personal information such as social security numbers or information that could permit access to financial accounts. While the law requires only notification, it is expected to impact cybersecurity because organizations are believed likely to prefer instituting improved security to disclosing breaches, with the latter's attendant reputation costs. It is also expected to have impacts beyond the state's borders, since interstate businesses are unlikely to institute separate cybersecurity procedures for different states.

Laws and regulations in other countries may also impact cybersecurity measures taken by organizations in the United States, especially if those organizations also engage in relevant activities in those countries. One example comes from the European Union (EU), which has adopted two directives that require organizations to implement cybersecurity measures. EU Directives 95/46/EC, on data protection, and 2002/58/EC, on privacy and electronic communications, require member nations to implement measures to ensure the protection of privacy of personal data held or communicated by organizations engaged in commercial or other relevant activities within the EU.

Partnerships

To varying degrees, critical infrastructure sectors are already involved in the development of cybersecurity frameworks. One way sector industries are working together is through voluntary partnerships called information sharing and analysis centers (ISACs).[200] DHS lists fourteen such centers.[201] The centers vary substantially in their activities and relationship with government.[202] In addition, some consortia have been formed to facilitate development and coordination of cybersecurity efforts. In addition to groups such as BITS, CIS, ISF, and ISSA mentioned earlier in this report, other examples include the Cybersecurity Industry Alliance (CSIA),[203] the Internet Security Alliance (ISA),[204] and the

National Cyber Security Partnership (NCSP).[205] There is significant variation, however, in the degree to which these groups are considered to be effective.

Working groups developed pursuant to the December 2003 National Cyber Security Summit developed reports with recommendations and guidance for improving cybersecurity. The Corporate Governance Task Force Report[206] recommended that organizations adopt governance measures[207] derived from ISO/IEC 17799, FISMA, and other sources. It also recommended that DHS endorse the recommendations and launch a public campaign urging their adoption by organizations. Other task forces produced reports with recommendations on software security, education and awareness for home users and small businesses, information sharing, and technical standards and the CC.[208]

POLICY OPTIONS

There is considerable public debate about whether efforts such as those described above are sufficient or if the federal government needs to take additional action to bolster cybersecurity in general and develop a national framework in particular. Supporters of stronger government efforts, including regulation, argue that they are necessary to improve security and will have a positive economic impact by reducing uncertainties concerning economic loss from cyberattacks. Opponents point to costs, the difficulty of determining what requirements are necessary and how to measure compliance, and problems in dealing with boundaries between networks and between nations.[209]

Some specific arguments that might favor legislative action are as follows:

- Most critical infrastructure is in private hands,[210] yet problems in these sectors arising from inadequate cybersecurity could have implications well beyond the sectors themselves.
- Some experts argue that cybersecurity is fundamentally a public good and therefore requires government involvement.[211] There are aspects of cyberspace that resemble those of a commons — an asset, such as a public road, that is generally available to the public rather than being in private hands. Unregulated commons can be susceptible to exploitation, degradation, and other problems.
- As the role of cyberspace in the U.S. and world economy continues to increase, its protection and reliability will become more clearly in the national interest, as is the case, for example, with commercial aviation and product safety.[212]

- The growing amount of personal information, including financial information, that is communicated via cyberspace makes it increasingly attractive to thieves and other criminals, making the law-enforcement function of government more relevant to cybersecurity.

The apparent failure of market incentives to stimulate adequate cybersecurity efforts means that governments may be required to intervene to correct the market failure. An often-cited example is the underreporting of security incidents by an organization, despite the fact that timely and sufficiently complete reporting is important for effective response as well as planning. Among the disincentives for reporting are damage to reputation[213] that can result in loss of customers or revenue, in the case of a company, or political repercussions in the case of a government agency. Laws such as the California Database Protection Act discussed above can help to correct such market failures, although they can themselves create market distortions or other problems.

However, many other observers believe that legislative action is unnecessary or inappropriate. Arguments against such action include the belief by many that regulation would be too intrusive and would stifle innovation, that voluntary efforts are sufficient, that time should be allowed for current laws and voluntary efforts to have impact before further legislative action is considered, and that the threat from cyberattack is not great enough to warrant further government action.

Models

Two models have sometimes been cited as providing possible avenues for federal efforts to develop a cybersecurity framework — the year-2000 (Y2K) computer problem, and environmental and safety regulations. Each is discussed below.

Response to the Year-2000 Computer Problem

Government involvement in the efforts to resolve the year-2000 computer problem[214] is sometimes cited as a possible model for government involvement in cybersecurity. A key element in that approach was the use of evolving and increasing requirements for publicly traded companies via actions by the Securities and Exchange Commission (SEC). The SEC promulgated rules to require companies to respond to the Y2K problem. Congress passed laws to facilitate information sharing and to reduce liability if the company had complied. This appeared to demonstrate that (1) the SEC can be effective in promoting changes such as those required for improved cybersecurity; (2) Congress can be an effective enabler of solutions — for example, by removing barriers to effective information sharing; and (3) a gradual, incremental approach can be effective.[215] However, critics respond that the model is inappropriate, for three reasons. First, the Y2K problem is thought by many to have been much less serious than feared, so that the effectiveness of response may be questionable as a reason for the low number of significant incidents. Second, the problem was fundamentally much simpler than cybersecurity, which may require a much more complex set of responses. Third, it was a one-time problem, whereas cybersecurity needs are continuous. Nevertheless, the lessons learned from the Y2K problem may usefully inform the cybersecurity response. In fact, the Sarbanes-Oxley Act uses the SEC to promote cybersecurity. Although the act does not directly provide for gradual improvement, the evolution of auditing standards could have that effect

Safety and Environmental Regulations

Whether cybersecurity standardization can be approached using safety or environmental regulation or similar efforts as a model might also be considered. For example, the analogy of cyberspace with the highway system (the "information superhighway") raises the question of whether governments might consider security regulations analogous to safety regulations that apply to roads, vehicles, and drivers.

Environment

The Environmental Protection Agency (EPA) administers several laws aimed at reducing and preventing environmental problems.[216] For example, the Pollution Prevention Act of 1990 required EPA "to develop and implement a strategy" for reducing pollution at the source.[217] In comparison, the HSA requires DHS

> to develop a comprehensive national plan for securing the key resources and critical infrastructure of the United States, including power production, generation, and distribution systems, information technology and telecommunications systems (including satellites), electronic financial and property record storage and transmission systems, emergency preparedness communications systems, and the physical and technological assets that support such systems.[218]

The plan would therefore include several key elements of cyberspace. However, the act does not explicitly provide DHS with authority to implement the plan.[219]

The Emergency Planning and Community Right-to-Know Act of 1986 requires many industrial facilities to report release of toxic chemicals annually. Submission of information on cybersecurity problems such as attacks remains voluntary under HSA, but the act contains provisions to facilitate such submissions for certain kinds of information.[220]

The Clean Air Act and related legislation requires EPA to set and enforce national standards for ambient air quality and assigns major responsibility for compliance to the states. It also requires standards for vehicle emissions that have led manufacturers to modify technologies to meet those standards. The HSA does not provide DHS with authority to set or enforce national standards for cybersecurity.

Food and Product Safety

Several federal agencies share responsibility for regulating food safety in the United States, primarily the Food and Drug Administration of the Department of Health and Human Services and the Food Safety and Inspection Service of the Department of Agriculture.[221] Both agencies set regulations with input from industry and other interested parties and use monitoring by inspectors to ensure conformance with the regulations. The Consumer Product Safety Commission is an independent regulatory agency established to protect the public from unreasonable risk of injury or death from consumer products. The agency sets voluntary and mandatory product safety standards, has the power to recall or even ban hazardous products.[222] Also, under state product liability laws, plaintiffs may sue for damages for injury to person or property resulting from a defective product.[223] The federal government does not regulate or set mandatory or voluntary standards for cybersecurity except to some extent with respect to federal agencies and contractors.

The federal approaches to environmental protection and to food and product safety could provide potential models should Congress wish to use government regulation as a way of improving cybersecurity, as some have proposed.[224] However, the highly interconnected, amorphous, and constantly evolving nature of cyberspace might provide significant barriers to the creation of regulations that improve cybersecurity but do not impede technology development and entrepreneurship.

Options for Congress

Should Congress consider taking action to facilitate the adoption of a framework for cybersecurity, there are several options that might be considered, with respect to both legislation and oversight. Some possibilities are described below, to illustrate the range of options. The examples are for illustration only.[225] Therefore, no discussion of benefits and disadvantages is given. Among the legislative options are the following.

Encourage the Widespread Adoption of Cybersecurity Standards and Best Practices

There are several potential ways to achieve such a goal. Perhaps the strongest measures would be for Congress to provide the Department of Homeland Security or another agency with regulatory authority over cyberspace industries and direct the department to develop and enforce mandatory cybersecurity standards, presumably through a process that involved the industries and other interested parties, as is the case, for example, with food safety. A moderate approach might be to further codify and strengthen DHS's role in working work with industry to develop voluntary standards and best practices, or to provide tax incentives for companies to adopt acceptable cybersecurity measures.

Leverage Procurement

Congress could require that cybersecurity be a high priority in all federal acquisitions of information technology. It could further require that companies that operated under accepted levels of cybersecurity practices, such as recognized international standards, would receive preference for federal contracts.

Encourage Mandatory Reporting

To further counter disincentives for companies to report cybersecurity vulnerabilities and breaches, Congress could require federal and private organizations to reveal certain kinds of security breaches, as California has done.[226] It could also require companies to report specified classes of incidents and vulnerabilities to DHS under the protections afforded by the HSA, rather than relying on voluntary reporting.

Facilitate Product Liability Actions

Congress could direct DHS to identify classes of cybersecurity weaknesses for which states could permit plaintiffs to sue manufacturers under product liability laws, or provide some other mechanism to facilitate such redress.

Facilitate Development of Cybersecurity Insurance

To the extent that lack of reliable actuarial information or other barriers are impeding the development of cybersecurity insurance, Congress could facilitate the development of the industry by providing for reinsurance or other guarantees, as it does in certain other areas.[227]

Strengthen Federal Cybersecurity Programs

H.R. 5068, introduced in the 108th Congress, and reintroduced in the 109th Congress as H.R. 285, proposes establishing the position of Assistant Secretary for Cybersecurity within DHS, with responsibility for, among other things, implementing priorities similar to those laid out in the NSSC. The 109th Congress might enact legislation with such provisions and also could strengthen existing cybersecurity efforts in other agencies such as NIST and NSF and through the SEC.

Oversight and investigative hearings could also provide mechanisms for Congress to facilitate the development of a cybersecurity framework. Several hearings were held during the 108th Congress, most notably by the Subcommittee on Cybersecurity, Science, and Research and Development of the House Select Committee on Homeland Security,[228] and the Subcommittee on Technology, Information Policy, Intergovernmental Relations and the Census of the House Committee on Government Reform.[229] Additional hearings could be held during the 109th Congress on legislative options such as those described above, or for example on implementation of the recommendations relating to vulnerability reduction in the *NSSC*.

Alternatively, Congress might decide that no legislative action or targeted oversight would be appropriate at this time. There remain considerable differences of opinion among experts about both the seriousness of cybersecurity threats, especially with respect to terrorism, and the potential benefits and disadvantages of any additional federal action to improve cybersecurity in the private sector. In addition, ongoing and new government and private sector efforts such as those discussed above might result in sufficient improvements to render additional congressional action unnecessary or of marginal benefit. The question remains, however, whether those efforts will have sufficient impact quickly enough to meet national cybersecurity needs. The answer to that question will likely depend to a significant degree on the scale and immediacy of cybersecurity threats and vulnerabilities, over which there is still considerable debate.

No matter what actions Congress might take to strengthen cybersecurity efforts, there are several issues that may be influential in the debate in addition to the uncertainties discussed above. One significant issue is the degree to which cyberspace is in fact a commons, with the attendant characteristics that reduce the likelihood that market mechanisms alone can lead to appropriate security. Another is the difficulty of obtaining coordination and cooperation from the large number of interested and affected parties, especially given the global nature of cyberspace. A third issue is whether the different approaches advocated by different groups can be effectively harmonized so that there is a common understanding of what the key elements of a cybersecurity framework should be and how they should be implemented. Related to that is the issue of how to ensure that the particular needs of different sectors are met while attempting to achieve harmonization. Finally, there is the issue of lag time — the degree to which the speed of evolution of cyberspace and its supporting technology outstrips attempts to develop effective standards and other elements of a cybersecurity framework. Those and other issues strongly suggest that the attempt to develop a national framework for cybersecurity is likely to remain a significant challenge for the nation during the 109th Congress.

References

[1] The White House, *National Strategy to Secure Cyberspace,* February 2003, [http://www.whitehouse.gov/pcipb/cyberspace_strategy.pdf].

[2] See, for example, F. William Connor and others, *Information Security Governance: A Call to Action,* Report of the Corporate Governance Task Force, April 2004, available at [http://www.cyberpartnership.org/init-governance.html]; and Chris Klaus and others, *Recommendations Report,* Report of the Technical Standards and Common Criteria Task Force, April 2004, available at [http://www.cyberpartnership.org/init-tech.html]. These reports discuss and examine frameworks within the scope of the issues each covers — governance and technical standards, respectively.

[3] Committee on National Security Systems (CNSS), National Security Agency, "National Information Assurance (IA) Glossary," CNSS Instruction No. 4009, May 2003, [http://www.nstissc.gov/Assets/pdf/4009.pdf]. p. 32. The glossary defines the 5 elements of IA as follows: *Authentication:* Security measure designed to establish the validity of a transmission, message, or originator, or a means of verifying an individual's authorization to receive specific categories of information (p. 4). *Availability:* Timely, reliable access to data and information services for authorized users (p. 5). *Confidentiality:* Assurance that information is not disclosed to unauthorized individuals, processes, or devices (p. 15). *Integrity:* Quality of an IS [information system] reflecting the logical correctness and reliability of the operating system; the logical completeness of the hardware and software implementing the protection mechanisms; and the consistency of the data structures and occurrence of the stored data. Note that, in a formal security mode, integrity is interpreted more narrowly to mean protection against unauthorized modification or destruction of

information (p. 34). *Nonrepudiation:* Assurance the sender of data is provided with proof of delivery and the recipient is provided with proof of the sender's identity, so neither can later deny having processed the data (p. 44).

[4] CNSS, "IA Glossary," defines *assurance* as a "measure of confidence that the security features, practices, procedures, and architecture of an IS [information system] accurately mediates and enforces the security policy" (p. 3).

[5] Thomas C. Glaessner and others, *Electronic Safety and Soundness: Securing Finance in a New Age,* World Bank Working Paper No. 26, (Washington, DC: The World Bank, February 2004), p. 9.

[6] "Cyberspace is composed of hundreds of thousands of interconnected computers, servers, routers, switches, and fiber optic cables that allow our critical infrastructures to work" (*NSSC* , p. vii). The *NSSC* also says that cyberspace refers to "…an interdependent network of critical information infrastructures…" (p. 13). These are somewhat narrower meanings than used in this report.

[7] One online definition is, "A metaphor for describing the non-physical terrain created by computer systems. Online systems, for example, create a cyberspace within which people can communicate with one another (via e-mail), do research, or simply window shop. Like physical space, cyberspace contains objects (files, mail messages, graphics, etc.) and different modes of transportation and delivery" (*Webopedia,* [http://www.webopedia.com/TERM/c/cyberspace.html], 21 March 2002. The term was coined in 1984 by science-fiction writer William Gibson, who apparently intended it, in his novel *Neuromancer,* to mean a form of virtual reality created by a world-wide set of interconnected computers and those who operated them.

[8] The term *virtual structure*, as used here, refers to the apparent or perceived organization or architecture created by or with information technology hardware and software.

[9] The bipartisan bill, which received no committee or floor action in the 108th Congress, further states, "(i) each of the terms 'damage' and 'computer' has the meaning that term has in section 1030 of title 18, United States Code; and (ii) each of the terms 'electronic communications system', 'electronic communication service', 'wire communication', and 'electronic communication' has the meaning that term has in section 2510 of title 18, United States Code."

[10] This difference can be seen in the IA Glossary definition of an information system as a "set of information resources organized for the collection,

storage, processing, maintenance, use, sharing, dissemination, disposition, display, or transmission of information" (CNSS, "IA Glossary," p. 33).

[11] An example of this usage is "...no cybersecurity plan can be impervious to concerted and intelligent attacks...", *NSSC,* p. 3.

[12] A *patch* is a piece of software code inserted into a computer program to fix a problem, or "bug," in the program.

[13] For example, "...increase the level of cybersecurity nationwide," *NSSC,* p. 2. This also appears to be the usage in H.R. 285.

[14] For example, "...programs to advance the training of cybersecurity professionals...," *NSSC,* p. 41.

[15] The government of the United Kingdom is concerned enough about this issue that it has launched a research program to discover ways to avoid catastrophic failures resulting from such "emergent properties" (Duncan Graham-Rowe, "Sprawling Systems Teeter on IT Chaos," *New Scientist,* 24 November 2004, [http://www.newscientist.com/news/ news.jsp?id=ns 99996706].

[16] See National Research Council, *Information Technology for Counterterrorism,* (Washington, DC: National Academy Press, 2003), p. 12-13 for a related discussion. The report characterizes attacks as potentially making a system or network unavailable, corrupted, or compromised.

[17] *Malware,* a contracted elision of *malicious software,* includes viruses, Trojan horses, worms, logic bombs, and any other computer code that has or is intended to have harmful effects.

[18] See CRS Report RL32331, *The Economic Impact of Cyber-Attacks.*

[19] See CRS Report RL31534, *Critical Infrastructure: Control Systems and the Terrorist Threat.* The acronym SCADA is derived from the term *Supervisory Control And Data Acquisition,* which refers to the function of those systems, which are often used to control processes in industrial facilities and to log information about status and conditions. They often communicate electronically with central computer systems that are connected to the Internet.

[20] For example, Martin Dodge and Rob Kitchin, *Atlas of Cyberspace* (Boston: Addison Wesley, 2001); or Martin Dodge, *An Atlas of Cyberspaces,* [http://www. cybergeography.org/atlas/atlas.html].

[21] See, for example, "Mapping the Internet," in Eric Fischer, Coordinator, *Understanding Cybersecurity,* CRS Workshop MM70048, July 21, 2003.

[22] After National Research Council, *Information Technology for Counterterrorism*, p. 12 — 13.

[23] For example, one definition of *risk* used by the National Institute of Standards and Technology (NIST) is "…a combination of: (i) the likelihood that a particular vulnerability in an agency information system will be either intentionally or unintentionally exploited by a particular threat resulting in a loss of confidentiality, integrity, or availability, and (ii) the potential impact or magnitude of harm that a loss of confidentiality, integrity, or availability will have on agency operations (including mission, functions, and public confidence in the agency), an agency's assets, or individuals (including privacy) should there be a threat exploitation of information system vulnerabilities," (National Institute of Standards and Technology, *Information Security, Volume II: Appendices to Guide for Mapping Types of Information and Information Systems to Security Categories,* NIST Special Publication 800- 60 Version 1.0, December 2003, p. 5). The same publication defines *threat* as "…any circumstance or event with the potential to intentionally or unintentionally exploit a specific vulnerability in an information system resulting in a loss of confidentiality, integrity, or availability," and *vulnerability* as "…a flaw or weakness in the design or implementation of an information system (including security procedures and security controls associated with the system) that could be intentionally or unintentionally exploited to adversely affect an agency's operations (including missions, functions, and public confidence in the agency), an agency's assets, or individuals (including privacy) through a loss of confidentiality, integrity, or availability" (p. 5).

[24] See, for example, National Institute of Standards and Technology, "Risk Management Guide for Information Technology Systems," NIST Special Publication 800-30, October 2001, p. 25.

[25] Cybercrime is usually distinguished from cyberterrorism just as crime is usually distinguished from terrorism, although the distinction is sometimes muddled in usage. Generally, they are distinguished based on the aim of the activity. *Cybercrimes* generally refers to crimes committed using information technology, especially the Internet, for personal gain, and *cyberterrorism* refers to crimes involving information technology that are performed for political ends.

[26] "…[A]n increased reliance on technology escalates the potential for, and the likelihood of, e-security threats. Furthermore, attacks … occur more often and with a polymorphic approach. Due largely in part to organized crime and terrorism, the speed and tenacity of the hacking community is growing at a rapid rate" (The World Bank, Integrator Group, "Global Dialogue 'Electronic Safety and Soundness," September 10, 2003, Summary, p. 2).

[27] For in-depth discussion of issues involved in the security of CI, see CRS Report RL30153, *Critical Infrastructures: Background, Policy, and Implementation.*

[28] 42 U.S.C. 5195c(e).

[29] The White House, *National Strategy for Homeland Security,* July 2002, p. 30, available at [http://www.whitehouse.gov/homeland/book].

[30] This categorization is used in the DHS/IAIP Daily Open Source Infrastructure Reports, available at [http://www.dhs.gov/dhspublic/interapp/editorial/editorial_0542.xml].

[31] See Shea, *Critical Infrastructure: Control Systems.*

[32] U.S.-Canada Power System Outage Task Force, *Interim Report: Causes of the August 14th Blackout in the United States and Canada*, November 2003, p. 30 — 31.

[33] Nuclear Regulatory Commission, "Potential Vulnerability of Plant Computer Network to Worm Infection," NRC Information Notice 2003-14, 29 August 2003, [ht tp: / /www.nrc.gov/reading-rm/doc-collections/gen-comm/info-notices /2003/in200314.pdf]; Kevin Poulsen, "Slammer Worm Crashed Ohio Nuke Plant Net," *The Register,* 20 August 2003. The infection was not deemed to pose a safety hazard.

[34] Tony Smith, "Hacker Jailed for Revenge Sewage Attacks," *The Register,* 31 October 2001, [http://www.theregister.co.uk/2001/10/31/hacker_jailed_for_revenge_sewage]. The perpetrator was a former employee of the company that had installed the system.

[35] Glaessner and others, *Electronic Safety and Soundness,* p. 10.

[36] For in-depth discussion of economic aspects of cybersecurity, see CRS Report RL32331.

[37] Glaessner and others, *Electronic Safety and Soundness,* p. 9.

[38] Timothy Allen, "WiFi Vulnerabilities," in Fischer, *Understanding Cybersecurity.*

[39] Election Data Services, "New Study Shows 50 Million Voters Will Use Electronic Voting Systems, 32 Million Still with Punch Cards in 2004," Press Release, 12 February 2004.

[40] For in-depth discussion of these issues, see CRS Report RL32139, *Election Reform and Electronic Voting Systems (DREs): Analysis of Security Issues.*

[41] A *worm* is a kind of malicious software, or *malware*, that can replicate and propagate itself without human assistance across a computer network, often causing harmful effects.

[42] Bruce Schneier, "Blaster and the Great Blackout," *Salon.com,* 16 December 2003, [http://www.salon.com/tech/feature/2003/12/16/blaster_security/index_np.html].

[43] There are various ways of hiding malware. A Trojan horse, for example, is malware disguised as something benign or useful. See Kenneth Thompson, "Reflections on Trusting Trust,@ *Communications of the ACM* 27 (1984): 761-763, available at [http://www.acm.org/classics/sep95]. He concluded that it can be essentially impossible to determine whether a piece of software is trustworthy by examining its source code, no matter how carefully. The entire system must be evaluated, and even then it can be very difficult to find malware.

[44] See, for example, Richard C. Linger and Carmen J. Trammell, "Cleanroom Software Engineering Reference Model, Version 1.0,@ Technical Report CMU/SEI-96-TR-022, November 1996, available at [http://www.sei.cmu.edu/pub/documents/96.reports/pdf/ tr022.96.pdf]

[45] "Open source software refers to a computer program whose source code is made available to the general public to be improved or modified as the user wishes" (CRS Report RL31627, *Computer Software and Open Source Issues: A Primer*, p. 1). What is "open" (or "closed") is the source code — what programmers actually write. This code is translated into machine code (compiled) for use by computers to run the programs. Machine code can be translated back into source code (decompiled). This does not recover the original source code but can be useful, for example, to hackers hoping to find vulnerabilities, or to defenders looking for malware that might be hidden in the machine code.

[46] National Security Agency (NSA), "Defense in Depth: A Practical Strategy for Achieving Information Assurance in Today's Highly Networked Environments," NSA Security Recommendation Guide, 8 June 2001, available at [http://nsa2.www.conxion.com/support/ guides/sd-1.pdf]. Sometimes these three elements are referred to as "people, process, and technology."

[47] These are defined in FISMA (44 U.S.C. 3542) as follows: *integrity:* "guarding against improper information modification or destruction, [including] ensuring information nonrepudiation and authenticity"; *confidentiality:* "preserving authorized restrictions on access and disclosure, including means for protecting personal privacy and proprietary information"; *availability:* "ensuring timely and reliable access to and use of information." Others define these terms somewhat differently (see, for

example, National Research Council, *Trust in Cyberspace,* (Washington, DC: National Academy Press, 1999), p. 301,303,307; and footnote 3 above), but the meaning is similar.

[48] Glaessner and others, *Electronic Safety and Soundness,* p. 45, in reference to ISO/IEC 13335, "Information Technology — Security Techniques — Guidelines for the Management of IT Security (GMITS)."

[49] NRC, *Trust,* p. 316. The report further defines trustworthiness as encompassing "correctness, reliability, security (conventionally including secrecy, confidentiality, integrity, and availability), privacy, safety, and survivability" (p. 14).

[50] NIST is attempting to do this for federal systems, as required by law (40 U.S.C. 11331 and 44 U.S.C. 3533). See, for example, NIST, *Standards for Security.*

[51] *NSSC,* p. viii.

[52] The Chemicals Sector Cyber-Security Information Sharing Forum Cyber-Security Strategy Task Team, *U.S. Chemicals Sector Cyber-Security Strategy,* June 2002, p. 8.

[53] The others address national security, homeland security, combating terrorism, combating weapons of mass destruction, physical protection of critical infrastructure and key assets, and money laundering. For a comparison, see General Accounting Office, *Evaluation of Selected Characteristics in National Strategies Related to Terrorism,* GAO-04-408T, 3 February 2004.

[54] NSA, "Defense in Depth." An in-depth discussion can be found in National Security Agency, "Information Assurance Technical Framework Document, Release 3.1," September 2002, available at [http://www.iatf.net/framework_docs/version-3_1/index.cfm].

[55] These are based in part on characteristics discussed in GAO, *Evaluation.*

[56] National Institute of Standards and Technology, *Generally Accepted Principles and Practices for Securing Information Technology Systems,* SP 800-14, September 1996. See also NIST, *Engineering Principles for Information Technology Security (A Baseline for Achieving Security),* SP 800-27, June 2001 (currently being revised).

[57] *NSSC,* p. 14 — 15.

[58] The Chemicals Sector Cyber-Security Information Sharing Forum, Cyber-Security Strategy Task Team, "U.S. Chemicals Sector Cyber-Security Strategy," June 2002.

[59] *NSSC,* p. 13.

[60] This and subsequent policy levels are after NIST, *Generally Accepted Principles,* p. 13 — 15.

[61] Glaessner and others, *Electronic Safety and Soundness,* p. 59.

[62] For example, one kind of attack involves sending victims email purportedly from a legitimate financial or software company and urging them to visit a website, also purportedly of this company, where they are requested to enter information such as a usernames and passwords for accounts. The hacker can then use this information to take control of the victim's computer or to steal funds.

[63] Karen Fogerty, "Chief Security Officers Lack Confidence in the State of their Organization's Information Security Efforts," *csoonline.com,* press release, 26 January 2004.

[64] See Cashell, *Economic Impact,* for further discussion of this issue.

[65] In a survey of chief security officers in early 2004, a majority assessed that their organizations were investing less than optimal amounts in security and that they were at best "somewhat confident" in the effectiveness of their cybersecurity activities. The survey also found that those investing more in cybersecurity had fewer incidents (Fogerty, "Chief Security Officers Lack Confidence").

[66] NSSC, p. 41.

[67] Fogerty, "Chief Security Officers Lack Confidence".

[68] For example, a summary of several reports by the National Research Council states, "From an operational standpoint, cybersecurity today is far worse than what known best practices can provide," (National Research Council, *Cybersecurity Today and Tomorrow: Pay Now or Pay Later,* (Washington DC: National Academy Press, 2002 p. 8).

[69] For example, sectors vary substantially in the degree to which they report cybersecurity incidents and their capital and operating expenditures on cybersecurity, both of which are thought to be associated with the level of commitment to effective security of those responsible for corporate governance — see Lawrence A. Gordon and others, *2004 CSI/FBI Computer Crime and Security Survey,* June 2004, available at [http://www.gocsi.com].

[70] See, for example, Corporate Governance Task Force, *Information Security Governance: A Call to Action,* April 2004, available from the National Cyber Security Partnership at [http://www.cyberpartnership.org/init-governance.html].

[71] Lee Rainie, "Half of Americans Fear Terrorists Might Mount Successful Cyber-Attacks against Key American Utilities and Businesses," Press Release, Pew Internet & American Life Project, 31 August 2003.

[72] One source is a 2002 survey by the National Cyber Security Alliance, which found that most home users were aware of the importance of Internet security but few followed recommended security practices such as updating virus definitions, using safer passwords, or installing and using firewall programs (Keith Nahigian, "Survey Gives Computer Users 'A' For Effort in Security Awareness, but Failing Grades for Follow-Through on Implementing Computer Safety Tools," Press Release, National Cyber Security Alliance, 17 September 2002). A more recent survey and scan done by the same group, in conjunction with America Online (AOL), found that most respondents reported that their home computers had been infected with a virus at least once, but that only one-third of those that currently had a virus (19%) were aware that they did; 80% had spyware or adware on their computers, but only 10% were aware of what they had; and that two-thirds did not have antivirus software that was regularly updated (America Online and the National Cyber Security Alliance, "AOL/NCSA Online Safety Study," October 2004, available at [http://www.staysafeonline.info/news/safety_study_v04.pdf]).

[73] NRC, *Cybersecurity Today and Tomorrow,* p. 8.

[74] For example, malware may be surreptitiously planted on home computers to turn them into components of a "bot net," where the computer is used, along with many others, to launch Internet attacks without the computer owner's knowledge, even while the owner is using the computer (Robert Lemos and Jim Hu, "'Zombie' PCs caused Web outage, Akamai says," *CNET News.com,* 16 June 2004).

[75] This is known as *reputation risk* (Glaessner and others, *Electronic Safety and Soundness,* p. 14.

[76] "e-security is more a reactive than a proactive practice, but this approach should be altered in order to decrease future threats" (Glaessner and others, *Electronic Safety and Soundness,* p. 26).

[77] For example, taking steps to minimize the disruptive impacts of a cyberattack would reduce its attractiveness to terrorists.

[78] Garrett Hardin, "The Tragedy of the Commons," *Science,* 162(1968):1243 — 1248.

[79] Glaessner and others, *Electronic Safety and Soundness,* p. 26.

[80] National Standards Policy Advisory Committee, "National Policy on Standards in for the United States and a Recommended Implementation Plan," December 1978, p. 6.

[81] This discussion is after NIST, "The ABC's of Standards-Related Activities in the United States," NBSIR 87-3576, May 1987, available at [http://ts.nist.gov/ts/htdocs/210/ncsci/ stdpmr.htm].

[82] Section 102(g) of the Homeland Security Act of 2002 requires that all DHS standards activities be "...be conducted in accordance with section 12(d) of the National Technology Transfer Advancement Act of 1995 (15 U.S.C. 272 note) and Office of Management and Budget Circular A — 119." The 1995 act requires federal agencies to use voluntary consensus standards except where they would be "inconsistent with applicable law or otherwise impractical." Circular A-119 provides guidance for implementing this provision.

[83] Some appear to consider the term to be essentially synonymous with "voluntary consensus standards." Others believe that it should embrace such additional concepts as "open use," which essentially means use without royalties or licensing restrictions (see Ken Krechmer, "The Principles of Open Standards," *Standards Engineering*, 50 (6) (November/December 1 9 98) , p . 1- 6 , available at [http://www.csrstds.com/openstds.html].

[84] National Institute of Standards and Technology, "About Standards Services Division (SSD)," 11 November 2002, [http://ts.nist.gov/ts/htdocs/210/ about.htm], and linked pages.

[85] American National Standards Institute, "About ANSI," n.d., [http://www.ansi.org/ about_ansi/overview/overview.aspx].

[86] "Memorandum of Understanding between the American National Standards Institute (ANSI) and the National Institute of Standards and Technology (NIST)," 27 December 2000, available at [http://ts.nist.gov/ts/htdocs/ 210/nttaa/ansimou.htm].

[87] National Institute of Standards and Technology, "Process Control Security Requirements Forum (PCSRF)," 26 January 2005, [http://www.isd.mel. nist.gov/projects/processcontrol].

[88] INCITS [http://www.incits.org/index.html] has a committee on security techniques which serves as the U.S. Technical Advisory Group to the Subcommittee on Security Techniques of the Joint Technical Committee on Information Technology of the International Organization for Standardization (ISO) and the International Electrotechnical Commission (IEC).

[89] IEEE engages in a broad range of standards activities [http://standards.ieee.org/sa/ index.html]; it has an Information Assurance Standards Committee (IASC) [http://ieeeia.org/ iasc/] and Task Force (TFIA) [http://ieee-tfia.org] involved in the development of various cybersecurity-related standards.

[90] [https://www.trustedcomputinggroup.org/home].

[91] [http://www.ietf.org/].

[92] [http://www.iso.org].

[93] [http://www.iec.ch].

[94] [http://www.jtc1.org].

[95] [http://www2.ni.din.de/sixcms/detail.php?id=10172].

[96] For example, 49 standards have been published under the direct responsibility of JTC1/SC27 (see [http://www.iso.org/iso/en/ stds development/tc/tclist/ TechnicalCommitteeStandardsListPage.TechnicalCommitteeStandardsList? COMMID=143] for a list).

[97] In the United States, these criteria were developed in the 1970s and 1980s in what came to be known as the "Orange Book." Europe developed its own criteria in the 1990s. The Common Criteria was developed in response to the market restrictions and other problems caused by having more than one set of criteria that security products would be required to meet (Kevin Hayes, "Common Criteria — A World Wide Choice," The Encyclopedia of Computer Security, 1998, available at [http://www.itsecurity.com/ papers/88.htm]. To evaluate conformance of products to the CC, NIST and NSA have developed a joint program, the National Information Assurance Partnership Common Criteria Evaluation and Validation Scheme (CCEVS) [http://niap.nist.gov/cc-scheme].

[98] Syntegra, "Common Criteria: An Introduction," n.d., available at [http://www. commoncriteriaportal.org/public/files/ccintroduction.pdf].

[99] Syntegra, "Common Criteria: User Guide," October 1999, available at [http://www. commoncriteriaportal.org/public/files/ccusersguide.pdf].

[100] National Security Agency, Information Assurance Directorate, "U.S. Government Firewall Protection Profile, for Medium Robustness Environments," 28 October 2003, available at [http://www.common criteriaportal.org/public/files/ppfiles/pp_vid1016-pp.pdf].

[101] Called a *target of evaluation* (TOE), this includes the product or system plus associated documentation.

[102] This is called a *security target.*

[103] A list of evaluated products is available from the Common Criteria Project at [http://www.commoncriteriaportal.org].

[104] "EALl is applicable where some confidence in correct operation is required, but the threats to security are not viewed as serious....EAL4...is applicable in those circumstances where developers or users require a moderate to high level of independently assured security in conventional commodity TOEs, and there is willingness to incur some additional security specific engineering costs," (Syngenta, "CC: An Introduction," p. 12 — 13). The highest level that the CC defines is EAL7, but there are no reports of evaluations above EAL4.

[105] National Research Council, *Trust in Cyberspace,* p. 199.

[106] See, for example, Information Security Forum, *The Standard, of Good Practice for Information Security,* March 2003, p. 5, available at [http://www.securityforum.org].

[107] JTC1 SC27, "Catalogue of SC27 Projects and Standards," SC27 Standing Document 7, 20 August 2003, [http://www2.ni.din.de/sixcms/media.php/1377/sc27n3647_sd7_catalog_ proj_stand_aug2003.htm]. Revision of Part 1 begun in 2000, other parts in subsequent years. The standard is being renamed as "Management of Information and Communications Technology Security" (MICTS) and the technical reports will become part of the standard.

[108] JTC1 SC27, "Catalogue."

[109] NIST, "International Standard ISO/IEC 17799:2000 Code of Practice for Information Security Management: Frequently Asked Questions," November 2002, available at [http://csrc.nist.gov/publications/secpubs/otherpubs/reviso-faq-110502.pdf].

[110] Sarah D. Scalet, "Guiding Lite," CSO Magazine, March 2003, available at [http://www.csoonline.com/read/030103/lite.html].

[111] JTC1 SC27, "Catalogue."

[112] Julie Kenward, "The Global Development of BS7799," The Encyclopedia of Computer Security, 2000, available at [http://www.itsecurity.com/papers/88.htm].

[113] NIST, "ISO/IEC 17799: Frequently Asked Questions." However, see below.

[114] This has been described as using the word "should" where a certifiable standard would use "shall." (Scalet, "Guiding Lite").

[115] For example, this was the approach reportedly taken by The Vanguard Group when it adopted the standard (Scalet, "Guiding Lite).

[116] See Scalet, "Guiding Lite."

[117] One organization that provides BS7799 certification within the United States in BSI Management Systems — USA, part of BSI Group, which is the publisher of BS7799.

[118] A *control* is defined in ISO/IEC 13335 as “a practice, procedure, or mechanism that reduces risk”; it may also be called a *safeguard* (C.J. Mitchell, “SC 27 Standing Document 6(SD 6), Glossary of IT Security Terminology (SC 27 N 2776),” JTC1 SC27, 31 March 2002, available at [http://www2.ni.din.de/sixcms/media.php/1377/sc27_standing_document_6_sc27n2776__terminology_.htm].

[119] Caroline Hamilton, “ISO-IEC 17799.The New International Standard for Information Security Management,” MS PowerPoint Presentation, May 2002, available at [http:// asisitsc.i8.com/library].

[120] ISF, based in Europe, is an international association of private companies and government organizations that performs research and provides information on cybersecurity to its members [http://www.securityforum.org].

[121] See [http://www.isfsecuritystandard.com/index_ie.htm].

[122] ITGI was founded by the Information Systems Audit and Control Association and its affiliated foundation in 1998 and provides information and research on information technology management [http://www.itgi.org].

[123] IT Governance Institute, *COBIT Mapping: Overview of International IT Guidance,* (Rolling Meadows, IL: ITGI, 2004), p. 5, available at [http://www.itgi.org].

[124] Ibid., p. 8.

[125] For example, under the process, “ensure systems security,” there are 21 specific control objectives, such as “Management should ensure that reaccredidation of security (e.g., through ‘tiger teams’) is periodically performed to keep up-to-date the formally approved security level and the acceptance of residual risk” (IT Governance Institute, “Control Objectives,” *COBIT,* 3rd Ed. (Rolling Meadows, IL: ITGI, June 2000), p. 102, available at [http://www.itgi.org]).

[126] ITGI, *COBIT Mapping*, p. 11 — 12.

[127] Specifically, the Federal Information Security Management Act of 2002 (FISMA), P.L. 107-347.

[128] National Institute of Standards and Technology, “Guideline for Identifying an Information System as a National Security System,” NIST Special Publication 800-59, August 2003.

[129] See NIST, “Federal Information Processing Standards,” 3 August 2004, [http://csrc. nist.gov/publications/fips/index.html].

[130] 15 USC 278g-3(a)(3).

[131] 5 USC 551 et seq. For a discussion of this an other federal management laws, see CRS Report RL30795, *General Management Laws: A Compendium*.

[132] Special publications present, in the "800 series," "documents of general interest to the computer security community" (Computer Security Response Center (CSRC), National Institute of Standards and Technology, "NIST Special Publications," [http://csrc.nist.gov/ publications/nistpubs/ index.html], 19 August 2004).

[133] NIST, *Generally Accepted Principles and Practices*.

[134] See CSRC, "NIST Special Publications."

[135] [http://www.iatf.net].

[136] National Security Agency, "Information Assurance Technical Framework (IATF) document, Release 3.1," September 2002, available at [http://www.iatf.net/framework_ docs/version-3_1/index.cfm].

[137] National Information Assurance Partnership Common Criteria Evaluation and Validation Scheme (CCEVS) Validation Body, "The Interpretations Process," 6 December 2004, [http://niap.nist.gov/cc-scheme/interps-process.html].

[138] See, for example, Subcommittee on Technology, Information Policy, Intergovernmental Relations and the Census House Committee on Government Reform, "Exploring Common Criteria: Can It Ensure that the Federal Government Gets Needed Security in Software?" Hearing, 17 September 2003, [http://reform.house.gov/TIPRC/Hearings/EventSingle.aspx? EventID=527].

[139] ANSI, "Standards Activities Overview," [http://www.ansi.org/ standards_activities/overview/overview.aspx], accessed 29 September 2004.

[140] ANSI, "ANSI Essential Requirements: Due Process Requirements for American National Standards," 30 January 2004, available at [http://www.ansi.org].

[141] Internet Engineering Task Force, "The Tao of IETF: A Novice's Guide to the Internet Engineering Task Force," RFC 3160, August 2001, [http://www.ietf.org/tao.html].

[142] This is how it is used in the context of the Common Criteria (see Syntegra, *Common Criteria for Information Technology Security Evaluation: User Guide,* October 1999, [http://niap.nist.gov/cc-scheme/cc_docs/ cc_users_guide.pdf]).

[143] This is also called security authorization (National Institute of Standards and Technology, *Guide for the Security Certification and Accreditation of Federal Information Systems,* NIST Special Publication 800-37, May 2004).

[144] BITS, "BITS Product Certification Program," [http://www.bitsinfo.org/fslab.html], 8 June 2004. The acronym BITS was derived from "Banking Industry Technology Secretariat."

[145] Gamma Secure Systems Ltd., "The Future of 7799," [http://www.gammassl.co.uk/bs7799/future.html], 7 July 2004.

[146] ISMS International User Group, "Certificate Register," [http://www.xisec.com/ register.htm], 23 July 2004.

[147] [http://www.isaca.org/].

[148] [https://www.isc2.org].

[149] See also the section below on measuring success.

[150] A national survey on computer crime and security conducted for the last several years has found little change in reporting of incidents by organizations experiencing intrusions, with about half of all participating organizations responding that they did not report them, with most of those citing concerns about reputation risk as a primary reason for not reporting (Gordon, *2004 CSI/FBI Survey,* p. 13).

[151] See, for example, NIST, *Generally Accepted Principles and Practices.*

[152] See NRIC, "NRIC Best Practices," [http://www.bell-labs.com/user/krauscher/nric/], 13 September 2004.

[153] ASP Industry Consortium, "A White Paper on Network Security for the ASP Industry," 2002; ASP Industry Consortium, "A White Paper on Platform Security for the ASP Industry," 2002. ASPs are application service providers, companies that use the Internet or other networks to provide other organizations with software-based services such as orderhandling. See [http://www.aspstreet.com] for information about the consortium.

[154] Carnegie-Mellon Software Engineering Institute, "Concept of Operations for the CMMI," 15 January 2001, [http://www.sei.cmu.edu/cmmi/background/conops.html].

[155] See, for example, Richard C. Linger and Carmen J. Trammell, "Cleanroom Software Engineering Reference Model, Version 1.0," Technical Report CMU/SEI-96-TR-022, November 1996, available at [http://www.sei.cmu.edu/pub/documents/96.reports/pdf/ tr022.96.pdf]).

[156] Information Systems Security Association, *Generally Accepted Information Security Principles,* Version 3.0 (2004), p. 2. GAISP is a successor to an earlier effort called Generally Accepted System Security Principles.

[157] See, for example, Federal Accounting Standards Advisory Board, "Generally Accepted Accounting Principles," [http://www.fasab.gov/ accepted.html].

[158] The document was released in July 2003 and is available at [http://www.bis. org/publ/ bcbs98.htm].

[159] Among them are authentication, nonrepudiation, segregation of duties, authorization, data access controls, encryption, recovery, intrusion detection, protection of data integrity, and incident response procedures.

[160] Alan Paller and Clint Kreitner, "Consensus Minimum Security Benchmarks," *IANewsletter,* 5, no. 3 (2002): 4 — 5, 9.

[161] Center for Internet Security, "Benchmarks/Tools," [http://www.cisecurity. org/ bench.html], n.d.

[162] These are available online through the Center for Internet Security at [http://www.cisecurity.org], for private-sector checklists, and through NIST at [http://csrc.nist.gov/ pcig/cig.html] for the federal government.

[163] This program is required by Sec 8(c) of the Cyber Security Research and Development Act of 2002, P.L. 107-305. It is known as the Security Configuration Checklists Program for IT Products (see [http://checklists.nist.gov]).

[164] [http://www.disa.mil].

[165] See Information Systems Security Association, "Welcome to the Security Management Index," [https://www.humanfirewall.org/smi/].

[166] One federal law with such requirements is the Sarbanes-Oxley Act of 2002, P.L. 107- 204.

[167] For more information, see the SAS-70 website, [http://www.sas 70.com/index2.htm].

[168] See, for example, General Accounting Office, *Federal Information System Controls Audit Manual,* GAO/AIMD-12.19.6, January 1999, available at [http://www.gao.gov/special.pubs/ ai12.19.6.pdf]; and National State Auditors Association and General Accounting Office, *Management Planning Guide for Information Systems S e c u r i t y A u d i t i n g ,* 10 De cember 2001, available at [http://www.nasact.org/techupdates/ downloads/ GAO/12_01-Mgmt_ Plan.pdf].

[169] See, for example, Information Systems Audit and Control Association, *IS Standards, Guidelines and Procedures for Auditing and Control Professionals,* 1 July 2004, available at [http://www.isaca.org/ Content Management/ContentDisplay.cfm?ContentID=13927]. This document defines *standards* as mandatory auditing and reporting requirements,

guidelines as guidance in applying the standards, and *procedures* as methods an auditor might use in an audit (p. 6).

[170] See, for example, Thomas Glaessner and others, Electronic Security: Risk Mitigation In Financial Transactions, The World Bank, June 2002, available at [http://www1.worldbank.org/finance/index.html]: "In many countries throughout the world, statistical analysis reveals that more than 50 percent of electronic security intrusions are carried out by insiders. An uneducated or undereducated workforce is inherently more vulnerable to this type of incident or attack. In contrast, a well-trained workforce, conscious of security issues, can add a layer of protection. Hence, the safety and efficiency of technology is directly related to the training and technical education of the persons using the technology" (p. 51).

[171] NSSC, p. 37 — 42.

[172] See [http://www.staysafeonline.info/]. The major government agencies supporting this alliance are DHS and the Federal Trade Commission.

[173] See CRS Report RL31846, *Science and Technology Policy: Issues for the 108th Congress, 2nd Session*, p. 16

[174] See [http://www.whitehouse.gov/omb/egov/a-1-fea.html].

[175] See, for example, Network Applications Consortium, "Enterprise Security Architecture: A Framework and Template for Policy-Driven Security," 3 December 2004, available at [http://www.netapps.org].

[176] See, for example, Gary Stoneburner and others, *Risk Management Guide for Information Technology Systems,* NIST Special Publication 800-30, July 2002, [http://csrc.nist.gov/publications/nistpubs/800-30/sp800-30.pdf], or CRA Reports, *Security Risk Management,* 2003, [http://www.foundstone.com/resources/whitepapers/wp_security_ risk_management.pdf].

[177] This is because cost-benefit analyses do not usually take externalities into account. Of course, even in the absence of direct impact, accepting such risks might nevertheless involve reputation costs.

[178] Of course, the incidence of viruses, Trojan horses, and other kinds of malware has increased steadily, as have attempts to compromise computers with them. Nevertheless, relatively simple measures can guard against most such attacks.

[179] Ronda Henning and others, *Workshop on Information System Security Scoring and Ranking,* 21 — 23 May 2001, Proceedings, (Silver Spring, MD: Applied Computer Security Associates, 2002).

[180] Marianne Swanson and others, *Security Metrics Guide for Information Technology Systems, NIST* Special Publication 800-55, July 2003, available at [http://csrc.nist.gov/publications/nistpubs/800-55/sp800-55.pdf].

[181] The guidelines provide the following example: "The impact metrics would quantify incidents by type (e.g., root compromise, password compromise, malicious code, denial of service) and correlate the incident data to the percentage of trained users and system administrators to measure the impact of training on security" (Ibid, p. 12).

[182] Glaessner and others, *Electronic Safety and Soundness,* p. 18 — 19.

[183] For a discussion of this and other cost issues in cybersecurity, see CRS Report RL32331 *The Economic Impact of Cyber-Attacks*.

[184] For example, computer failure was a significant factor in the August 2003 electrical blackout in the northeastern United States (U.S.-Canada Power System Outage Task Force, "Final Report on the August 14, 2003 Blackout in the United States and Canada: Causes and Recommendations," April 2004, [https://reports.energy.gov/BlackoutFinal-Web.pdf]). The economic cost of that blackout has been estimated at several billion dollars, with most of that loss occurring outside the electric utility sector (Electricity Consumers Resource Council, "The Economic Impacts of the August 2003 Blackout, " 9 February 2004, [http://www.elcon.org/ Documents/ Economic Impacts OfAugust2003Blackout.pdf]).

[185] For example, collective concern regarding reputation risk or the potential for government intervention might lead organizations within a sector to agree on a minimum standard of cybersecurity practice.

[186] *NSSC,* p. ix. C

[187] Ibid., p. 13.

[188] For a discussion of specific federal provisions with respect to both government and private systems, see CRS Report RL32357, *Computer Security: A Summary of Selected Federal Laws, Executive Orders, and Presidential Directives*.

[189] See National Institute of Standards and Technology, "FISMA Implementation Project," [http://csrc.nist.gov/sec-cert/index.html], 2 November 2004.

[190] Office of Management and Budget, "FY2003 Report to Congress on Federal Government Information Security Management," 1 March 2004, [http://www.whitehouse.gov/omb/inforeg/fy03_fisma_report.pdf]. The report does not include information on national security systems.

[191] Government Accountability Office, "Information Security: Continued Efforts Needed to Sustain Progress in Implementing Statutory Requirements," Statement before the Subcommittee on Technology, Information Policy, Intergovernmental Relations and the Census, House

Committee on Government Reform, GAO-04-483T, 16 March 2004, [http://www.gao.gov/new.items/d04483t.pdf], p. 4.

[192] See, for example, National Security Agency, "Security Configuration Guides," [http://www.nsa.gov/snac/], n.d.

[193] For details, see Moteff, *Computer Security.*

[194] The auditing standard for internal controls released in June 2004 by the Securities and Exchange Commission contains numerous references to information technology controls (Public Company Accounting Oversight Board, *Auditing Standard No. 2 — An Audit of Internal Control Over Financial Reporting Performed in Conjunction with An Audit of Financial Statements,* 9 March 2004, [http://www.pcaobus.org/Rules_of_the_Board/Documents/Rules_of_the_Board/Auditing_Standard_2.pdf]).

[195] Dawn Kawamoto, "Hidden Gold in Corporate Cleanup," CNET News.com, 24 November 2004, [http://news.com.com/2102-1029_3-5465305.html].

[196] For example, a service company that is not directly covered by these acts but that provides relevant services to companies that are will likely be expected by its customers to institute appropriate cybersecurity measures.

[197] Sec. 404(a)(1).

[198] Office of the Inspector General, Department of Homeland Security, *Progress and Challenges in Securing the Nation's Cyberspace,* OIG-04-29, July 2004, available at [http:// www.dhs.gov/interweb/assetlibrary/OIG_CyberspaceRpt_Jul04.pdf].

[199] [http://www.us-cert.gov].

[200] ISACs were established pursuant to language in Presidential Decision Directive/NSC-63, "Critical Infrastructure Protection," 22 May 1988, directing the federal government to encourage their creation. Known as PDD-63, it was superseded by Homeland Security Presidential Directive/HSPD-7, "Critical Infrastructure Identification, Prioritization, and Protection," 17 December 2003.

[201] They include food, water, fire services, law enforcement, state government, information technology, telecommunications, research and education, electric power, energy, surface transportation, financial services, chemical industry, and real estate (see Department of Homeland Security, "Threats and Protection: Information Sharing and Analysis Centers," 13 January 2003, [http://www.dhs.gov/dhspublic/display?theme=73&content=1375].

[202] See, for example, CRS Report RL32597, *Information Sharing for Homeland Security: A Brief Overview.*

[203] CSIA is a consortium of cybersecurity companies ([https://www.csialliance.org]).

[204] ISA is a collaborative effort of Carnegie Mellon University and the Electronic Industries Alliance ([http://www.isalliance.org]).

[205] NCSP is a public-private partnership involving industry, government, and academia ([http://www.cyberpartnership.org]).

[206] F. William Connor and others, *Information Security Governance: A Call to Action,* Report of the Corporate Governance Task Force, April 2004, available at [http://www.cyberpartnership.org/init-governance.html].

[207] The report calls these measures an "information security governance framework."

[208] These reports are available through the NCSP at [http://www.cyberpartnership.org].

[209] Jeffrey E. Payne, "Regulation and Information Security," *IEEE Security & Privacy* (March/April 2004): 32-35.

[210] The White House, *National Strategy for the Physical Protection of Critical Infrastructures and Key Assets,* February 2003, [http://www.whitehouse.gov/pcipb/physical.html].

[211] Glaessner and others, *Electronic Safety and Soundness,* p. 7.

[212] "[T]his is an area where legislation and regulations are necessary. It is a proper government responsibility to require cyber-security upgrades. It is as important as other consumer protections, such as food and product safety. "Accordingly, virtually everyone accepts a broad but controlled safety regulation of standards at meat packers, auto manufacturers or financial institutions. We're all OK with government mandating certain shields to keep our skies safe. Establishing minimum security standards in these and other key areas should be looked at in the same light" ("Government has role in fighting cyber terrorism," *San Francisco Business Times,* 2 December 2002, available at [http://sanfrancisco.bizjournals.com/sanfrancisco/stories/2002/12/02/ editorial3.html]).

[213] Glaessner and others, *Electronic Safety and Soundness,* p. 18.

[214] Basically, this was a problem with the way most computer software had been designed to handle dates. Much software code was designed when computer memory was at a premium and therefore was coded to process only the last 2 digits of the year. Therefore, the program could not distinguish an entry for the year 2000 from one for the year 1900. This could be particularly a concern for programs that used clocks for core processes. Predictions about the impacts of this discrepancy ranged from trivial to disastrous effects. Because of the risks of a disastrous impact, a substantial effort was launched to modify or replace computer code (both programmable and hardwired) to correct the problem.

[215] Payne, "Regulation and Information Security, " p. 35.

[216] For more detail about these statutes, see CRS Report RL30798, *Environmental Laws: Summaries of Statutes Administered by the Environmental Protection Agency.*

[217] 42 USC 13103.

[218] Sec. 201(d)(5).

[219] Also, the *NSSC* was not developed pursuant to any specific legislative mandate, and no federal agency has been given statutory authority to implement it.

[220] Specifically, the act defines a protected class of nonpublic "critical infrastructure information" relating to security (for example, instances of attacks or known vulnerabilities) and prohibits public disclosure and certain other use by DHS of such information if voluntarily submitted to the federal government.

[221] See CRS 98-91, *Food Safety Agencies and Authorities: A Primer*; and CRS Report RL31853, *Food Safety Issues in the 109th Congress.*

[222] [http://www.cpsc.gov].

[223] See CRS Issue Brief IB97056, *Products Liability: A Legal Overview.*

[224] See, for example, Christ Strohm, "Tenet warns of terrorists combining physical, telecommunications attacks," GovExec.com, 1 December 2004, [http://www.govexec.com/ dailyfed/1204/120104c1.htm].

[225] CRS does not take positions on legislative issues.

[226] See discussion of the California Database Protection Act under the section above on current efforts.

[227] For example, in the wake of the September 11, 2001 attacks, Congress enacted the Terrorism Risk Insurance Act (P.L. 107-297), to assist both policyholders and the insurance industry in adjusting to the impacts of the attacks.

[228] For a list of hearings and other activities, see Subcommittee on Cybersecurity, Science, and Research & Development of the U. S. House of Representatives Select Committee on Homeland Security, "Cybersecurity for the Homeland," Report of the Activities and Findings by the Chairman and Ranking Member, December 2004, available at [http://hsc.house.gov/ files/cybersecurityreport12.06.0 4.pdf].

[229] A list of subcommittee hearings in the 108th Congress and links to testimony are available on the committee website at [http://reform. house.gov/TIPRC/Hearings/?Timeframe= Past&CatagoryID=117].

INDEX

A

access, 5, 7, 11, 15, 18, 32, 40, 52, 61, 66, 76
accidents, 7
accountability, 18, 19, 21, 24
accounting, 43
acquisitions, 57
activities, 2, 5, 6, 7, 9, 10, 12, 13, 22, 29, 34, 46, 52, 68, 70, 71, 74, 81
actuarial, 45, 58
adaptability, 22, 40, 42
adaptation, 21
adjustment, 45
administration, 20
Administration, 56
Administrative Procedure Act, 34
administrators, 78
adware, 69
agriculture, 14
alternative, 32, 36
alternatives, 39, 40
ambient air, 56
America Online, 69
amorphous, 9, 56
anthropogenic, 14
antivirus software, 69
application, 2, 15, 21, 31, 33, 36, 40, 42, 75
arrest, 47
Asia, 38
assessment, 3, 11, 24, 32, 34, 39, 45
assets, 6, 14, 22, 34, 50, 55, 64, 67
assignment, 18
association, 37, 41, 73
Atlas, 63
attacks, 1, 9, 10, 12, 13, 15, 16, 18, 19, 23, 25, 27, 28, 39, 46, 56, 63, 64, 69, 77, 81
attention, 1, 10, 12, 13, 14, 27, 46
attractiveness, 69
auditing, 28, 43, 55, 76, 79
Australia, 15
authentication, 6, 7, 30, 76
authenticity, 19, 66
authority, 37, 41, 51, 56, 57, 81
availability, 6, 7, 19, 50, 64, 66, 67
aviation, 53
awareness, 32, 35, 41, 44, 49, 53

B

banking, 14
barriers, 45, 49, 55, 56, 58
Basel Committee, 41
behavior, 19, 23, 47
bell, 75
benchmark, 42
benchmarks, 2, 28, 33, 42, 46
benefits, 35, 36, 57, 58
benign, 66
Best Practice, 39, 57, 75

best practices, 2, 3, 28, 29, 39, 40, 41, 42, 57, 68
bipartisan, 62
Boston, 63
breaches, 17, 23, 39, 45, 52, 57
Britain, 33
British, 32
buildings, 10, 11, 13
Bush, 20
Bush administration, 20
Bush, President George W., 20
business, 1, 10, 15, 32, 33, 36, 44, 45, 52

C

cables, 11, 62
California, 52, 54, 57, 81
Canada, 65, 78
capital, 35, 68
cast, 16
casting, 16
categorization, 65
cattle, 28
cellular phone, 11
Census, 58, 74, 78
certification, 18, 28, 33, 34, 35, 36, 38, 39, 40, 42, 43, 44, 73
certifications, 38, 39
channels, 10
chemical, 13, 14, 15, 22, 79
chemical industry, 15, 79
chemicals, 21, 56
CIS, 42, 52
civil liberties, 22
classes, 9, 40, 51, 57
classification, 28, 32
classified, 29
Clean Air Act, 56
clients, 45
commerce, 2
commercial, 52, 53
commitment, 18, 24, 68
Committee on Homeland Security, 58, 81
commodity, 72
commons, 28, 45, 47, 53, 59
communication, 7, 9, 29, 62
communication systems, 9
community, 28, 64, 74
competency, 37
complexity, 2, 5, 9, 10, 16, 27
compliance, 22, 32, 36, 39, 50, 53, 56
components, 2, 5, 6, 7, 9, 10, 11, 12, 13, 14, 16, 19, 21, 22, 23, 27, 28, 41, 44, 49, 69
computer, 1, 2, 5, 6, 7, 11, 13, 15, 16, 21, 23, 25, 31, 33, 34, 50, 51, 54, 62, 63, 65, 66, 68, 69, 74, 75, 78, 80
computer software, 16, 80
computer systems, 1, 15, 16, 23, 50, 62, 63
computer use, 25
computers, 6, 7, 10, 11, 16, 42, 62, 66, 69, 77
computing, 11, 19, 30
Conference Committee, 29
confidence, 6, 25, 32, 38, 62, 64, 72
confidentiality, 6, 7, 19, 50, 64, 66, 67
configuration, 42
confusion, 43
Congress, 3, 6, 50, 54, 56, 57, 58, 62, 77, 78, 81
connectivity, 19
consensus, 2, 29, 33, 37, 39, 40, 42, 70
consolidation, 1
consumer protection, 80
consumers, 17, 30
content, 39, 79
continuing, 38
continuity, 32, 42, 51
contractors, 51, 56
contracts, 57
control, 6, 10, 11, 13, 14, 15, 17, 18, 23, 29, 32, 33, 51, 63, 68, 73
controlled, 15, 80
coordination, 49, 52, 59
copyright, 5
corporate governance, 51, 68
corporations, 13, 17, 20, 44
corruption, 9
cost-benefit analysis, 45
cost-effective, 21, 39, 51
costs, 10, 28, 39, 44, 45, 46, 47, 49, 52, 53, 72, 77

countermeasures, 12
credit, 10, 15
credit card, 10, 15
crime, 2, 10, 12, 64, 75
crimes, 15, 64
criminals, 1, 7, 13, 15, 54
critical infrastructure, vii, 2, 13, 14, 16, 19, 22, 50, 52, 53, 55, 62, 67, 81
CRS, 63, 65, 66, 74, 77, 78, 79, 81
cryptographic, 34
culture, 41
customers, 47, 51, 54, 79
cyber terrorism, 80
cyberattack, 9, 10, 15, 25, 54, 69
cybercrime, 12, 15, 25
cybersecurity, 1, 2, 3, 5, 6, 7, 8, 9, 10, 11, 12, 13, 14, 16, 17, 18, 20, 21, 22, 23, 24, 25, 27, 28, 29, 30, 31, 32, 34, 35, 36, 39, 40, 41, 42, 43, 44, 45, 46, 47, 49, 50, 51, 52, 53, 54, 55, 56, 57, 58, 63, 65, 68, 71, 73, 78, 79
cyberspace, 2, 5, 6, 7, 8, 9, 10, 11, 12, 13, 14, 16, 27, 28, 32, 35, 41, 44, 45, 47, 51, 53, 54, 55, 56, 57, 59, 61, 62
cyberterrorism, 12, 64

D

data processing, 6
data structure, 61
death, 56
defense, 1, 14, 20, 22, 35, 45, 46
definition, 5, 7, 28, 62, 64
degradation, 28, 53
degree, 24, 25, 40, 42, 53, 58, 68
deliberation, 40
delivery, 46, 62
demand, 47
denial, 9, 11, 78
Department of Agriculture, 56
Department of Defense, 31, 40
Department of Health and Human Services, 56
Department of Homeland Security, viii, 1, 6, 57, 79
dependent, 14
destruction, 5, 9, 10, 14, 61, 66, 67
detection, 6, 7, 12, 22, 46, 76
development, 1, 3, 12, 17, 18, 28, 29, 30, 32, 33, 34, 35, 36, 39, 40, 41, 44, 49, 51, 52, 56, 58, 71
DHS, 14, 24, 29, 51, 52, 53, 55, 56, 57, 58, 65, 70, 77, 81
directives, 52
disabled, 15
disclosure, 5, 42, 66, 81
disposition, 63
distortions, 46, 54
distribution, 55
diversity, 2, 27
division, 51
DoS, 9, 10
draft, 37
due process, 37
duration, 22, 50
duties, 76

E

EA, 44
economic, 7, 12, 14, 25, 28, 46, 47, 53, 65, 78
economic incentives, 28, 46
economic security, 14
economy, 2, 22, 50, 53
education, viii, 3, 16, 24, 25, 28, 38, 44, 53, 77, 79
Education, 44
efficiency, 46, 77
election, 16
electric power, 79
electric utilities, 15
electrical, 15, 37, 78
electricity, 9, 14
electronic, 2, 6, 7, 9, 11, 15, 30, 31, 52, 55, 62, 77
electronic communications, 7, 52, 62
electronic systems, 6
email, 6, 16, 22, 68
e-mail, 62

Emergency Planning and Community Right-to-Know Act, 56
emergency preparedness, 55
emergency response, 51
employees, 23
encryption, 22, 30, 76
energy, 13, 14, 15, 78, 79
engineering, 3, 17, 23, 40, 72
Enhancement, 6
enterprise, 22, 28, 36, 44
Enterprise, 44, 77
entrepreneurship, 56
environment, 2, 6, 22, 24, 31, 41, 43
environmental, 19, 32, 54, 55, 56
Environmental Protection Agency, 55, 81
EPA, 55, 56
ethical, 41
Europe, 71, 73
European, 30, 52
European Union (EU), 52
evaluation, 30, 31, 38, 45, 71
evidence, 24, 25, 39, 47, 49
evolution, 29, 35, 41, 55, 59
execution, 46
Executive Order, 78
expectations, 21, 23, 47
expenditures, 36, 68
expertise, 33
experts, 1, 16, 17, 25, 30, 33, 37, 39, 40, 41, 53, 58
exploitation, 53, 64
externalities, 47, 77
extortion, 12

F

failure, 2, 15, 24, 47, 54, 78
FBI, 68, 75
February, 61, 62, 65, 67, 78, 80
federal, 1, 3, 5, 6, 14, 20, 29, 30, 34, 35, 42, 44, 46, 49, 50, 51, 53, 54, 56, 57, 58, 67, 70, 74, 76, 78, 79, 81
federal government, 44, 50, 51, 53, 56, 76, 79, 81
federal law, 6, 29, 34, 50, 51, 76
Federal Trade Commission, 77
fiber, 62
fidelity, 15
finance, 12, 14, 77
financial institutions, 15, 51, 80
financial markets, 10
fire, 79
firewalls, 7, 31
firm, 43
fish, 28
flexibility, 21, 33, 36, 40, 42
flow, 6
focusing, 18, 27, 34
food, 14, 56, 57, 79, 80
Food and Drug Administration, 56
food safety, 56, 57
foreign, 10
framework, 1, 2, 3, 5, 7, 9, 10, 11, 12, 13, 18, 19, 20, 21, 30, 31, 33, 35, 39, 41, 43, 49, 50, 51, 53, 54, 57, 58, 59, 67, 74, 80
fraud, 12, 15, 25
free, 38
fund transfers, 31
funds, 68

G

GAAP, 41
Gamma, 75
GAO, 67, 76, 79
General Accounting Office, 67, 76
Generally Accepted Accounting Principles, 41, 76
generation, 15, 16, 55
global economy, 22
goals, 2, 3, 18, 19, 20, 21, 22, 23, 29, 39
Gold, 79
governance, 13, 17, 18, 22, 23, 24, 33, 34, 41, 50, 51, 53, 61, 68, 80
government, 1, 3, 5, 9, 10, 14, 15, 21, 22, 24, 30, 37, 43, 44, 47, 49, 50, 51, 52, 53, 54, 55, 56, 58, 63, 73, 77, 78, 79, 80
Government, 49, 50, 54, 58, 71, 74, 78, 80
Government Accountability Office, 50, 78
Government Accountability Office (GAO), 50

government intervention, 78
government involvement, 53, 54
Grades, 69
Great Britain, 33
groups, 12, 20, 52, 53, 59
guidance, 22, 23, 32, 35, 36, 41, 42, 53, 70, 77
guidelines, 2, 6, 21, 28, 31, 32, 34, 41, 42, 46, 50, 51, 77, 78

H

hackers, 12, 15, 17, 47, 66
hacking, 64
handling, 32
hands, 53
harm, 11, 64
harmful, 63, 65
harmful effects, 63, 65
harmonization, 22, 59
health, 14, 51
health information, 51
Health Insurance Portability and Accountability Act, 51
high risk, 12, 27
high-level, 20, 32
hips, 31
Homeland Security, 1, 6, 14, 57, 65, 70, 79
Homeland Security Act, 1, 70
homes, 16
House, 58, 61, 65, 74, 78, 80, 81
House Committee on Government Reform, 58, 74, 79
households, 1
housing, 13
human, 19, 22, 50, 65
humans, 7

I

id, 63, 71
identity, 10, 62
IETF, 37, 74
implementation, 17, 18, 19, 32, 34, 40, 42, 46, 49, 58, 64
incentive, 47, 49
incentives, 3, 19, 20, 23, 27, 28, 47, 50, 51, 54, 57
incidence, 39, 77
income, 47
indication, 27
industrial, 11, 14, 29, 56, 63
industry, 14, 15, 20, 24, 29, 34, 35, 38, 40, 56, 57, 58, 79, 80, 81
inefficiency, 47
infection, 15, 65
Information, 1, 5, 15, 21, 29, 30, 32, 33, 34, 35, 38, 41, 42, 43, 50, 51, 58, 61, 63, 64, 65, 66, 67, 68, 70, 71, 72, 73, 74, 75, 76, 77, 78, 79, 80, 81
information processing, 41
information sharing, 52, 53, 54
Information System, 21, 29, 38, 41, 42, 64, 73, 75, 76, 77
information technology, 5, 7, 15, 27, 29, 30, 33, 44, 55, 57, 62, 64, 73, 79
infrastructure, 1, 2, 6, 7, 9, 10, 11, 13, 14, 16, 22, 29, 32, 35, 50, 52, 53, 55, 67, 81
Infrastructure, 1, 14, 29, 51, 63, 65, 79
infrastructures, 19, 22, 50, 62
injury, 56
innovation, 54
input, 56
Inspection, 56
Inspector General, 79
inspectors, 56
institutions, 15, 16
insurance, 3, 45, 58, 81
Insurance, 51, 58, 81
insurance companies, 45
integration, 22
integrity, 6, 7, 19, 28, 50, 61, 64, 66, 67, 76
interest, 15, 53, 74
interface, 29
internal controls, 51, 79
international, 1, 20, 22, 30, 41, 57, 73
International Organization for Standardization, 30, 70
International Organization for Standardization (ISO), 30, 70

international standards, 30, 57
Internet, 1, 2, 9, 11, 13, 14, 30, 37, 42, 52, 63, 64, 69, 74, 75, 76
Internet Engineering Task Force, 30, 37, 74
interoperability, 44
interstate, 52
intervention, 78
intrusions, 23, 75, 77
investigative, 58
investment, 3, 24, 36, 47
ISO, 30, 31, 32, 33, 34, 35, 36, 38, 39, 41, 43, 53, 67, 72, 73
issues, 2, 31, 32, 34, 51, 58, 61, 65, 77, 78, 81

J

January, 15, 68, 70, 74, 75, 76, 79

L

language, 35, 79
law, 5, 29, 47, 52, 54, 67, 70, 79
law enforcement, 47, 79
laws, 50, 51, 52, 54, 55, 56, 57, 74
lead, 12, 37, 45, 46, 49, 59, 78
leadership, 18, 24, 42, 44
learning, 16
legal, 23, 42, 43, 49, 50
legislation, 2, 56, 57, 58, 80
legislative, 7, 53, 54, 57, 58, 81
licensing, 70
likelihood, 59, 64
limitations, 21, 22, 46
links, 81
literature, 39
local government, 15, 24
losses, 15

M

machines, 11, 16
maintenance, 17, 19, 32, 63
malicious, 63, 65, 78
malware, 10, 16, 17, 65, 66, 69, 77
Management, 5, 41, 43, 45, 50, 64, 67, 70, 72, 73, 74, 76, 77, 78
mapping, 11
market, 3, 21, 46, 47, 49, 51, 54, 59, 71
market failure, 3, 47, 54
market incentives, 54
markets, 10
mathematical, 40
meanings, 2, 5, 6, 11, 21, 62
measurement, 29, 31
measures, 3, 6, 7, 12, 15, 18, 25, 46, 52, 53, 57, 77, 79, 80
meat, 80
media, 11, 72, 73
memorandum of understanding, 29
memory, 80
messages, 62
metaphor, 6, 62
methodology, 21, 39, 41
metric, 39, 46
Microsoft, 16, 42
military, 5
missions, 64
Model, 66, 75
models, 40, 54, 56
money, 12, 67
money laundering, 12, 67
monitoring, 15, 32, 34, 56

N

nation, vii, 2, 7, 14, 19, 59
national, 1, 2, 5, 7, 9, 10, 13, 14, 19, 20, 21, 22, 29, 30, 34, 44, 49, 50, 53, 55, 56, 58, 59, 67, 75, 78
National Institute of Standards and Technology(NIST), 64, 67, 70, 73, 74, 75, 78
National Research Council, 63, 67, 68, 72
national security, 22, 34, 50, 67, 78
National Strategy, 1, 14, 61, 65, 80
National Strategy to Secure Cyberspace, 1, 61
natural, 14
network, 5, 6, 15, 28, 30, 40, 62, 63, 65
networking, 15

networks, 6, 7, 9, 11, 16, 23, 33, 53, 75
NIST, 21, 22, 29, 30, 34, 42, 46, 50, 51, 58, 64, 67, 68, 70, 71, 72, 73, 74, 75, 76, 77
nongovernmental, 22, 44, 49, 50, 51
nongovernmental organization, 22, 44, 50
Northeast, 15
NRC, 65, 67, 69
nuclear power, 15
nuclear power plant, 15
Nuclear Regulatory Commission, 65

O

objectives, 19, 20, 23, 31, 32, 34, 43, 44, 73
obsolete, 20, 27
OECD, see Organization for Economic Cooperation and Development, 41
Office of Management and Budget, 50, 70, 78
Ohio, 65
oil, 15
oil refineries, 15
OMB, 50
online, 37, 62, 76
operating system, 2, 13, 16, 31, 42, 61
operator, 19
organization, 9, 17, 18, 19, 20, 21, 22, 23, 24, 25, 29, 31, 32, 33, 39, 40, 41, 42, 43, 44, 45, 54, 62, 73
Organization for Economic Cooperation and Development (OECD), 41
organizations, 1, 9, 13, 15, 17, 20, 21, 22, 23, 24, 25, 28, 29, 30, 32, 34, 35, 36, 38, 39, 40, 42, 43, 44, 46, 50, 51, 52, 53, 57, 68, 73, 75, 78
organized crime, 12, 64
oversight, 42, 57, 58

P

Paper, 62, 75
participation, 22, 30, 37
partnership, 40, 44, 80
Partnership, 53, 68, 71, 74
partnerships, 22, 47, 50, 52
password, 78
peers, 47
penalties, 22
perception, 13, 25
performance, 19, 29, 31, 33, 42
periodic, 21, 22
permit, 16, 52, 57
personal, 10, 15, 51, 52, 54, 64, 66
phone, 11
planning, 21, 32, 34, 35, 44, 54
plants, 20
play, 3, 10, 13
policies, 2, 6, 13, 18, 22, 23, 32, 39, 43, 45, 46, 49, 50
policy, 3, 22, 23, 24, 32, 46, 49, 50, 62, 68
policy levels, 68
policymakers, 10
political, 7, 54, 64
Pollution Prevention Act, 55
poor, 28, 44, 46
power, 15, 19, 55, 56, 79
Power, 65, 78
preference, 21, 57
premium, 80
preparedness, 25, 55
prevention, 7, 12, 19
preventive, 25, 27, 28
preventive approach, 28
primary, 36, 75
priorities, 20, 44, 58
privacy, 21, 22, 42, 52, 64, 66, 67
private, 1, 2, 3, 9, 13, 15, 20, 22, 24, 29, 34, 40, 42, 44, 47, 49, 50, 51, 52, 53, 57, 58, 73, 76, 78, 80
private sector, 1, 3, 9, 20, 34, 49, 51, 58
private-sector, 2, 13, 15, 24, 29, 49, 50, 51, 76
proactive, 69
probability, 12
procedures, 2, 7, 13, 17, 18, 23, 24, 29, 34, 37, 39, 40, 43, 46, 52, 62, 64, 76, 77
process control, 20, 29
processes, 1, 6, 7, 14, 17, 18, 23, 34, 38, 40, 43, 46, 61, 63, 80
production, 14, 55
profession, 41

proficiency testing, 46
profit, 25
program, 22, 32, 34, 38, 42, 44, 46, 50, 63, 66, 71, 76, 80
programs, 3, 13, 16, 32, 44, 46, 50, 63, 66, 69, 80
promote, 3, 44, 55
property, 55, 56
protection, 6, 7, 10, 19, 22, 30, 31, 34, 42, 52, 53, 61, 67, 76, 77
protocols, 6, 13
proxy, 46
public, 1, 2, 9, 13, 14, 22, 25, 28, 37, 40, 44, 47, 50, 52, 53, 56, 64, 66, 71, 80, 81
public awareness, 13, 25, 50
Public Company Accounting Oversight Board, 79
public health, 14
public safety, 2
punitive, 47

R

R&D, see research and development, 52
race, 27
radio, 15
range, 10, 15, 30, 31, 49, 50, 57, 71
reading, 65
real estate, 79
recall, 56
recognition, 30
recovery, 7, 12, 19, 28, 51, 76
reduction, 19, 45, 58
reengineering, 44
regulation, 3, 21, 29, 36, 41, 47, 50, 53, 54, 55, 56, 80
regulations, 52, 54, 55, 56, 80
regulators, 36
reinsurance, 58
relationship, 19, 52
relevance, 13, 40
reliability, 19, 53, 61, 67
Reliability, 40
report, 2, 3, 6, 8, 10, 11, 23, 24, 33, 34, 43, 45, 50, 52, 56, 57, 62, 63, 67, 68, 75, 78, 80
reputation, 15, 36, 47, 52, 54, 69, 75, 77, 78
research, 7, 16, 52, 62, 63, 73, 79
Research and Development, 58, 76
resolution, 37, 49
resource allocation, 23
resources, 21, 49, 51, 55, 62, 77
responding, 21, 75
responsibilities, 7, 18, 21, 22, 24
responsibility, 18, 34, 41, 42, 56, 58, 71, 80
responsiveness, 41
restoration, 6, 7
returns, 24
revenue, 54
risk, 6, 11, 12, 13, 14, 15, 22, 23, 25, 27, 28, 32, 40, 41, 42, 45, 50, 56, 64, 69, 73, 75, 77, 78
risk assessment, 15, 32, 41, 45
risk management, 28, 45
risks, 2, 12, 23, 25, 32, 38, 41, 45, 46, 47, 77, 80
rivers, 15
roadmap, 20
royalties, 70

S

safeguard, 73
safeguards, 32
safety, 2, 14, 37, 53, 54, 55, 56, 57, 65, 67, 69, 77, 80
Sarbanes-Oxley Act, 43, 51, 55, 76
SAS, 43, 76
scalable, 31
science, 62
Science and Technology Directorate, 52
SEC, 54, 58
Secretary of Defense, 5
Securities and Exchange Commission (SEC), 54
security, 1, 3, 5, 6, 7, 8, 13, 14, 15, 16, 17, 18, 19, 20, 21, 22, 23, 24, 25, 28, 29, 30, 31, 32, 33, 34, 35, 38, 39, 40, 41, 42, 43, 44, 45, 46, 50, 51, 52, 53, 54, 55, 57, 59, 61, 62, 64, 65, 66, 67, 68, 69, 70, 71, 72, 73, 74, 75, 77, 78, 80, 81

segregation, 76
self, 34, 41, 43
self-assessment, 34, 43
September 11, 2001, 81
series, 34, 74
service, 9, 11, 14, 16, 42, 75, 78, 79
service provider, 75
services, 3, 6, 7, 13, 14, 15, 16, 22, 23, 29, 32, 34, 38, 40, 41, 44, 46, 50, 61, 75, 79
sewage, 15, 65
sharing, 55, 63
shipping, 14
skills, 15
social, 14, 23, 52
social impacts, 14
social security, 52
software, 3, 6, 7, 11, 12, 16, 17, 18, 30, 31, 40, 42, 43, 53, 61, 62, 63, 65, 66, 68, 69, 75, 80
software code, 63, 80
solutions, 22, 32, 35, 55
specialization, 37
specificity, 22, 42
speed, 42, 59, 64
spyware, 69
staff, 33
stakeholders, 5, 7, 40
standardization, 29, 55
standards, 2, 3, 28, 29, 30, 31, 32, 33, 34, 35, 36, 37, 38, 39, 41, 42, 43, 50, 51, 53, 55, 56, 57, 59, 61, 70, 71, 74, 76, 80
state laws, 50
statistical analysis, 77
statutes, 81
statutory, 81
storage, 55, 63
strategic, 19, 21, 27
strategies, 18, 20, 22, 23, 32, 39
stress, 2, 7, 21, 42
study, 69
suffering, 47
Survey, 68, 69, 75
survivability, 67
systematic, 39
systems, 1, 2, 5, 6, 7, 9, 11, 13, 14, 15, 16, 17, 19, 22, 23, 28, 29, 30, 31, 32, 33, 34, 40, 41, 42, 43, 47, 50, 55, 62, 63, 67, 73, 78

T

tangible, 18
targets, 10, 12, 25
task force, 53
tax incentives, 57
technical assistance, 51
technological, 35, 55
technology, 2, 5, 7, 15, 17, 18, 20, 22, 23, 25, 27, 29, 30, 33, 35, 36, 41, 44, 46, 55, 56, 57, 59, 62, 64, 66, 73, 77, 79
telecommunications, 10, 11, 14, 55, 79, 81
telephone, 11, 15
terrorism, 20, 58, 64, 67, 80
terrorist, 1, 10, 25
terrorists, 1, 7, 10, 12, 13, 15, 25, 69, 81
Terrorists, 69
testimony, 81
The Homeland Security Act, 51
theft, 6, 10, 12, 13, 15, 16, 25
theory, 11, 41
threat, 11, 12, 22, 24, 41, 51, 54, 64
threats, 6, 7, 11, 25, 31, 45, 51, 58, 64, 69, 72
threshold, 10
tiger, 73
time, 19, 30, 36, 38, 44, 54, 55, 58, 59
title, 50, 62
top management, 22
toxic, 56
traffic, 11
training, 3, 7, 18, 24, 25, 28, 35, 44, 63, 77, 78
transaction costs, 49
transactions, 6, 9
transfer, 45
transmission, 11, 55, 61, 63
transportation, 9, 14, 15, 62, 79
Trojan horse, 63, 66, 77
trustworthiness, 19, 67

U

UK, see United Kingdom, 63
United Kingdom, 63
updating, 23, 69
US, 51
US, see United States, 1, 10, 14, 15, 16, 22, 24, 33, 38, 50, 51, 52, 55, 56, 62, 65, 70, 71, 73, 78
US-CERT, 51
users, 13, 15, 16, 28, 44, 53, 61, 69, 72, 74, 78

V

validity, 61
values, 41
vandalism, 16
variability, 50
variation, 24, 53
vehicles, 55
victims, 25, 68
virtual reality, 62
viruses, 27, 63, 69, 77
voiding, 45
voters, 15
voting, 15, 37
vulnerability, 11, 12, 13, 16, 19, 27, 44, 58, 64

W

Washington, 62, 63, 67, 68
waste, 15
water, 14, 15, 79
weakness, 2, 12, 13, 24, 49, 50, 64
weapons, 10, 67
weapons of mass destruction, 67
websites, 10
White, 61, 65, 75, 80
White House, 61, 65, 80
wireless, 15
work, 20, 44, 57, 62
workers, 24
workforce, 24, 77
World Bank, 62, 64, 77
worm, 15, 65
worms, 10, 16, 63

Y

yield, 9